Practical Guide of Home Gardening: A Complete Book

Practical Guide of Home Gardening: A Complete Book

Surinder Bhanot

Practical Guide of Home Gardening: A Complete Book

ISBN 978-93-5111-836-7

Published in 2016 in India by

RANDOM PUBLICATIONS

4376-A/4B, Gali Murari Lal, Ansari Road
New Delhi-110 002
Phone : +9111-43580356, 011-23289044, 011-43142548
e-mail: sales@randompublications.com,
info@randompublications.com, randomexports@gmail.com

Reprinted 2025

Type Setting by : Friends Media, Delhi-110089
Digitally Printed at: Replika Press Pvt. Ltd.

Preface

The purpose of the book is to help farmers improve family food supplies and nutrition year round through home gardening.

The home garden has a special significance because among a family's most important basic needs are food and shelter.

There are many household items which can deter insects. Sucking insects such as aphids can be deterred by sprinkling ash over the insects. They are usually on the underside of leaves. Ash sprinkled around the base of plants can deter some crawling insects. Soapy water poured or sprayed over sucking insects can also be effective. Slugs and other pests can be trapped in a hall: buried bottle with a little beer remaining in the bottom. Coffee grounds will deter many insects.

Certain plants are known to repel many types of insect, and some farmers plant these as companions to food crops. Garlic, marigold and lemon grass are some of these plants.

Some farmers know how to prepare natural pesticides from extracts of certain plants, seeds or fruit which can be mixed with water and sprayed on to plants. Some common examples in Southeast Asia are tobacco, neem fruits, rotenone and oil from citrus skin. In general, a farmer must experiment a little to find an effective solution which is easy to prepare. Do not forget that these natural pesticides can also be poisonous to animals and humans. Follow the same safety rules as with chemical pesticides.

Gardens, and the people who work them, help local agriculture to develop and diversify by exchanging knowledge of practices and technologies, trading seeds and animal breeds, and also by sharing knowledge and/or collaboration in marketing. Some skills and technologies learned and developed in a garden later support local agriculture. In a community with gardens there are always different crops and small livestock at different stages of production, resulting in a fairly regular supply of surplus garden produce for trade in local markets, and for supply of farm inputs such as seeds and livestock for growing on. A regular supply of local produce supports the viability of local markets, which in turn provide growers the sales outlets for income. People tend to trust and prefer local markets and food from local gardens and agriculture. Gardens enhance this interdependent system of local agriculture, markets, and local consumers.

This book will definitely prove to be a boon to teachers, students and research scholars.

– Author

Contents

1

Introduction to Garden

The term *garden,* which is of Germanic origin,means "yard" or "enclosure" and denotes ways of organizing earth,water,plants and, sometimes, people, animals, and art (sculpture,architecture,theater,music,and poetry),the formal qualitities of which are determined as much by pleasure,artistry,or aesthetics as by convenience or necessity.This definition excludes arrangements of sacred space based on religious customs and sports,exclusions that are consistent with most societies. Not all cultures have gardens.For many reasons,anthropologists and garden historians consider most small cultivated plots to be forms of agriculture,as opposed to gardens.Gardens presuppose agriculture but in addition embrace a cultural and psychological distance from agriculture expressed in aesthetics.

GARDENS IN THE HISTORY OF IDEAS

Gardens have the capability to give physical form to ideas either by being modeled on familiar ideas or by creating a new design that generates or evokes new ideas,or through a combination of the two.Gardens make abstract ideas concrete—visible,tangible,and kinesthetic.In so doing,gardens can communicate complex abstract ideas convincingly.

DEATH

Gardens express ideas of victory over death in three ways.First,since their living components could die at any time (as a result of neglect or the whim of the owner or overwhelming natural forces),their mere existence represents a triumph over ill-will,chaos,and death; gardens signal that the world can be made right,especially through the use of human knowledge,skill,and spirit.Second,because gardens' biological materials inevitably grow,die,decay,and are then reconstituted to form life once again,they provide a powerful symbol of the cyclical aspect of life,negating death's apparent finality with a metaphorical triumph over death,fear,and hopelessness.This biological cycle implicit in the garden suggests a transmutation of death and an antidote to despair.Finally,depending on the culture,the form gardens are given reflects either (a) their triumph over the chaos found in nature,a chaos that is perceived as a constant threat to humanity—and in monotheistic cultures,a symbol of humanity's distance from God,or (b) the tremendous power of nature,of which humanity is a necessary part.In this manner,the garden's form reflects the innate hope that humans express by either taming or cooperating with nature.This hope,this expression,allows beholders to feel part of larger forces,bigger than their own short lives and limited powers.These three symbolic triumphs over death and fear are so compelling that exceptions,such as the "monster" sculptures in one part of the Boboli Garden at the Pitti Palace,are rare.Usually they are ironic: the agony of Christ in the garden of Gethsemane; the tortures of damnation in Hieronymous Bosch's painting *The Garden of Earthly Delights.*

TIME AND TEMPORALITY

Related to death are the various ideas of time and temporality—the internal experiences of time.Gardens in seasonal climates reveal the cyclical experience of the seasons,and all gardens underscore the cycles of day and night.Through these cycles one becomes more aware of the passing of time,of recurrence and passing away forever.Gardens often utilize sundials,or poems,to highlight the awareness of a particular idea of time. Poems and allusions of all kinds,as well as relics and historical artifacts,can also be used to make people aware of the past (or some idea of it) and of their collective or personal histories.

ORDER AND PLENTY

The people of ancient Egypt understood that by controlling the Nile River and the agriculture dependent upon it,they might impose order on the primordial chaos that was always a potential threat.Egyptian garden paintings,the world's earliest,show geometry and symmetry as the formal indications of this valuable foundation of such order.These early images show rectangular pools filled with fish,ducks,and lotus surrounded by regularly spaced fruit trees—emblematic of an ideal of the good life as it exists around the world.

The idea of the garden as a place where order is imposed upon an inherently chaotic,disorderly,painful,and dangerous natural world is central to ancient Egyptian,Persian,Islamic,European,and European-American concepts of the garden.The noted landscape gardener Lancelot "Capability" Brown (1715–1783),famous for designs of gardens that looked liked natural landscape,considered his efforts as improvements on the natural state (as well as on the rigid and geometric designs of previous gardenists); even nineteenth-century Romantic-era gardens,which thrived on the appearance of disorder,were carefully planned. Related to the idea of an order that provides for humanity—and to the idea of the garden as a triumph over death—is the idea of the garden as a site of never-ending bounty,never failing with the seasons.This idea is more common in India and the monotheistic Middle East,Europe,and America.Homer described the garden of Alcinous,king of the Phaeceans,in *The*

Odyssey(book vii): "and verdant olives flourish round the year./The balmy spirit of the western gale/eternal breathes on fruits untaught to fail." Chinese and especially Japanese gardens differ in being more likely to celebrate the different beauties of the several seasons. European villa gardens,of both the informal "pastoral" and the more formal French types,reflected instantiate the notion of the garden as a place of plenty by extolling the ideal of a close relationship to agriculture.Often this closeness was literal: gardens were situated within the larger farm,and might include (geometric) herb gardens,grape arbors,or symmetrically planted fruit groves; adjacent fields were actively cultivated.

THE LOST HOME

Nebuchadnezzar (r.605–562 b.c.e.),the Chaldean king of Babylon,introduced another persistent idea of the garden,that of the garden as a lost home.Nebuchadnezzar built the Hanging Gardens,one of the Seven Wonders of the ancient world,to comfort one of his wives,who missed "the meadows of her mountains,the green and hilly landscape of her youth" (Thacker,p.16).A similar motivation prompted the creation of the Tuileries Gardens in Paris.The Qing emperor Kang Xi (1662–1723) built the Pi-shu shan-chuang at Rehe (Jehol) in China to emulate the Manchurian homelands.In the modern era,retirees in the deserts of the American Southwest,self-exiled from temperate climates,recreate the comforting lawns,maples,and flowers reminiscent of their previous homes.Homer used Odysseus's memories of his childhood in his garden with his father,who gave him fruit trees and taught him the names of the plants,to underscore the hero's longing for home.

The Hanging Gardens of Babylon were described by Diodorus Siculus:

Since the approach to the garden sloped like a hillside and the several parts of the structure rose from one another tier on tier ... [it] resembled a theater ... the uppermost gallery,which was fifty cubits high,bore the highest surface of the park ... the roofs of the galleries were covered with beams of stone ... sufficient for the roots of the largest trees; and the ground,when levelled off,was thickly planted with trees of every kind that ... could give pleasure to the beholder ... The galleries contained many royal lodgings; and there was one gallery which contained openings leading from the topmost surface and machines for supplying the garden with water.(Thacker,p.17)

Admired by the ancient Greeks and Romans,they demonstrate several characteristics of gardens persisting to the early twenty-first century: the use of engineering and technology—often,paradoxically,to achieve a natural effect—and the attempt to make the garden a place of pleasure and sensuous delight; the integration of agriculture in the garden; and the integration of theater,poetry,and painting. In terms of world history (not just garden design),the most far-reaching,if poignant,image of the garden as lost home is found in the story of the Garden of Eden in the *Book of Genesis.*

GARDEN AS PARADISE AND ENCLOSURE

In 401 b.c.e.,the Greek historian Xenophon,in his *Oeconomicus,* Book IV,introduced the idea of the pleasure garden (Persian,*paradeisos,* "enclosure") to Greece,based on gardens he had seen while fighting in Persia,and recommended its imitation.Xenophon's description of the Persian gardens was again popularized in 1692 by the Englishman William Temple in his influential*Essay upon the Gardens of Epicurs: Or,of Gardening,in the Year 1685* (1692):

a paradise among them [the Persians] seems to have been a large Space of Ground,adorned and beautified with all Sorts of Trees,both of Fruits and of Forest,either found there before it was inclosed [sic],or planted after; either cultivated like Gardens,for Shades and for Walks,with Fountains or Streams,and all Sorts of Plants usual in the Climate,and pleasant to the Eye,the Smell,or the Taste; or else employed,like our Parks,for Inclosure [sic] and Harbour of all Sorts of Wild Beasts,as well as for the Pleasure of Riding and Walking: And so they were of more or less Extent,and of differing Entertainment,according to the several Humours of the Princes that ordered and inclosed [sic] them.... (quoted in Hunt and Willis,pp.96–97)

Enclosure is central to many types of gardens,including ancient Roman courtyards,which were surrounded by the house,Chinese and Korean coutyard gardens,and Japanese dry rock gardens (*karesansui*).Unlike the medieval European cloister gardens derived from them,Roman gardens often included mural paintings of gods and landscapes.

The Persian model of the garden as a paradise on earth later evolved into the Islamic *chahar-bagh,* an enclosed quadrangular garden with central perpendicular paths or canals dividing it into four equal sections.Later made famous in carpets and brought to India by the Mughals,*chahar-bagh* s,the most famous example of which is the Taj Mahal,reached Europe as the medieval *hortus conclusus,* or enclosed garden,as a result of the Crusades (1050–1150) and through Islamic gardens in Spain,Italy,and Sicily.Many twentieth-century rose gardens continue this form.Like its Islamic prototypes,the medieval garden was practical and symbolic,evoking the earthly and spiritual pleasures of the biblical Paradise and the Garden of Eden.Secular poetry such as the medieval French *Le Roman de la Rose,* purportedly composed during a dream in a rose garden,extended the *hortus conclusus* to include romantic love and Platonic ideals of fulfilment.Paintings and tapestries,too,especially the unicorn tapestries at Cluny and the Cloisters of the Metropolitan Museum of Art,showing unicorns and the "Lady in the Garden," take on various symbolic and allegorical readings.

The Garden of Eden,of course,is also a "lost home." Eden is described in *Genesis* as a kind of*chahar - bagh,* enclosed,divided into quarters by rivers meeting at right angles in the center,containing the Tree of the Knowledge of Good and Evil.Construed by Jews,Christians,and Muslims as an actual place from the human past,it was distinguished from Paradise,which was an ideal realm to be experienced by the righteous or the beloved of God in the future,and for Christians after their death or after the Second Coming.Eden as a prototype for European gardens was vastly expanded by the literary version John Milton presented in his epic *Paradise Lost* (1667) as a natural landscape (Hunt,p.79).

The earliest East Asian depictions of the four Buddhist paradises show Buddhas and Bodhisattvas in palaces surrounded by fragrant trees and flowers,dancers,and musicians.As cave paintings at Dunhuang (China) show,the religious significance of lotus suggested using the visual image of the lotus floating in a pond; azure rectangles of water with pink blossoms began to appear,growing larger,eventually with buds showing souls reborn in paradise as babies.This inspired actual gardens,including the famous Buddhist garden at Anapchi,in Kyongju,Korea (674 c.e.).

The pond at Anapchi shows a second visual allusion to paradise,the Daoist Isles of the Immortals.Depicted here as actual islands,they are sometimes represented by rocks on an "ocean" of dry gravel.Originating with the Chinese emperor Wudi (141–87 b.c.e.),such depictions were originally intended to attract the Immortals themselves,in the hope they would share their secrets.

A number of Fujiwara-period (989–1185) Japanese gardens,starting with the Byodo-in's Phoenix Hall (a villa in Uji outside Kyoto),simulated the Mahayana Paradise described in a sutra that attested to women's ability to reach enlightenment (yiengpruksawan).These gardens were designed to make paradise tangible and imaginable.Since the daughters of the Fujiwara clan were consistently married off to emperors,their eventual enlightenment was important both to their families and to the nation.The ability to visualize paradise was believed to facilitate enlightenment (legendary Queen Vaidehi's instruction by Buddha in meditation through visualization,including visualization of Paradise,was painted at Dunhuang).The construction of gardens as an aid to such visualization meant that the empresses would actually attain Paradise more easily.

GARDEN AS RUSTIC RETREAT

The Romans invented the idea of the villa—a home and farm in the countryside—which they believed provided a manner of living superior to that of city life.The villa achieves this goal both physically and spiritually or culturally by affording self-expression,self-cultivation,and self-definition.The garden became the setting and the occasion of this ideal,where by emulating cultured and educated men one became more cultured and educated oneself.(Although the model is largely patriarchal,a few women have done the same: the Duchess Eleonora di Toledo [wife of Cosimo I] in the Boboli Garden,the Countess of Bedford at Moor-Park [Hertforshire],and Mildred Bliss at Dumbarton Oaks,designed by Beatrix Farrand).

The Roman pastoral ideal symbolized by the villa and its garden ideal was revived in Renaissance Italy and eighteenth-century England,whence it spread to America.It was based on models found in Vergil's *Aeneid* and *Georgics,* in the letters of Pliny the Younger,and in the architect Andrea Palladio's (1508–1580) books and buildings reinterpreting classical architecture.Pliny's celebratory descriptions of life in the country influenced literati in their creation of an image of a life worth living—and in their designs of gardens within which to live.The Greek and Roman forms of government presupposed politically active citizens,and the resuscitation of these models as an ideal form elicited from Europeans,Britons,and Americans active participation not only in governing but in reimagining the ways the world might be governed and how social intercourse could be encouraged; creating gardens as representations of these emerging worldviews was part of the process of re-imaging.

The idea of the villa garden as a realm of personal cultivation in which one emulates historical role models is strikingly similar to one set of East Asian ideas of gardens, wherein gardens serve as places for contemplation, scholarship, artistic engagement, and social interaction with other literati. From the time of Wang Wei (699–c.760),East Asian paintings represent gardens of the literati as places of retreat from the corrupt world of everyday affairs,places that made possible the personal cultivation or "self-transformation" according to Buddhist,Confucian,and Daoist models.Paintings of the scholar Tao Yuanming (365–457),famous for his integrity,show a rustic fence and a few chrysanthemums depicting the garden whose tending was the pretext of his retirement. Fruit-bearing trees and other forms of agriculture were important parts of Chinese villas and literati gardens of "retirement" (Clunas),a feature that also recalls Paradise and the notion of plenty. In both Europe and East Asia,villa gardens as ideal realms eliciting personal cultivation coexist with the ideal of the rustic retreat,be it a shell-lined grotto or a humble thatched hut.Both remained vital for centuries,inspiring garden construction and permitting endless reformulation of intellecutal literati ideas and ideals.

GARDEN AS ART

Gardens were often regarded as art,both in theory and as a result of intimate and intricate relations among gardens and other arts.Most famous is Horace Walpole's (1717–1797) theory of the interrelations of the "three arts," poetry,painting,and the garden,in his *History of the Modern Taste in Gardening* (1771–1780).

The inclusion of carved or handwritten poetic quotations is found frequently in Asian and European gardens of nearly all styles,and visiual allusions to well-known poems,legends,or stories provide the basis for garden vignettes,such as the flat angular "eight-plank" bridges alluding to the tales of Ise in Japanese gardens,as well as themes for garden "rooms" or motifs.Highly influential gardens have been designed—or described—by poets,most famously Murasaki Shikibu (d.c.1014–1020),John Milton,Alexander Pope (1688–1744),Ishikawa Jozan (1583–1672),and Yuan Mei (eighteenth century).

Italian,French,German,and British formal gardens were used for theatrical entertainments and masques,sometimes in pavilions designed for the purpose (later gardens included shallow amphitheaters),while masques,operas,and other forms of early modern theater often had scenes set in a garden. Architecturally,gardens encompass—or are encompassed by—a house,palace,or temple.But formal gardens,English "natural" landscape gardens,European-American romantic gardens,and large Chinese gardens also incorporated small pavilions or "follies"; Japanese gardens often feature small rustic tea houses,or halls (later donated to temples).All have bridges both decorative and useful.

GARDEN AS MICROCOSM OF NATURE

East Asian thinking in many cases centered on the nature of the cosmos,the relation of yin and yang,the place of human beings in nature,and so on.These ideas were at once conceptual/intellectual,artistic,spiritual,and experiential/ imaginative,designed to provide the scholar with an opportunity for contemplation of nature like that provided within the landscape itself.Miniature gardens assembled on trays (bonsai) presented the macrocosm in microcosm for the viewer to use to immerse himself in nature or contemplate the Dao.

Sixteenth-century European explorers returned home with exotic plants that were featured in the new scientific botanical gardens.The first botanic gardens were quartered geometric arrangements simulating the "four quarters of the world" with plants in the area allotted to their continent of origin.Twentieth-century botanic gardens specialized in the creation of miniature systems representing whole environments,either cultural (Japanese gardens) or biological,recreating specific biomes (tropical,desert) and capable of sustaining the plants (and sometimes animals) native to them.(The Denver and Brooklyn Botanic Gardens have both types.) In the twenty-first century zoos (formally known as "zoological gardens") have started re-creating the topography and native vegetation of the animals in their collections,thus becoming more like gardens.

GARDEN AS MICROCOSM OF THE STATE

In Han China (202 b.c.e.–220 c.e.) gardens such as Tu Yuan (Rabbit Park) were used to "extend the grandeur of the princely dwelling,to be a site for ceremonies and magic,and to continue the time-honored mold of a game park" (Morris,p.13).The contemporary Chinese imperial garden described by Pere Attiret (1757) was designed to represent the country for the emperor,whose status forbade traveling freely. From the sixteenth century,French gardens were used politically in myriad ways: "to impress foreigners with the power of the court,to stir the loyalty of Frenchmen and,after the political and religious crisis deepened in the second half of the sixteenth century,to subtly express the political policy of the state.The court festival,especially as it was masterminded by Catherine de Médicis,often provided an opportunity to bring together opposed factions,turning their 'real conflicts into a chivalrous pastime'" (Adams,p.33).Versailles has been shown to be an elaborate four-dimensional demonstration of the power of Louis XIV; decision-making that went into the planning of its park was explicitly political (Berger).The British designed and interpreted gardens that were symbolic of the state and of political power.Such a connection was first drawn in Britain by Shakespeare in the gardener's speech in *Richard II* (act 3,scene 4).

The propensity to use gardens to express political arguments and commitments,and to understand garden design in political terms,permitted the garden historian Walpole to associate French formal gardens with monarchy and tyranny,and "natural" growth and irregularity with the newly emerging opposition government (Chase; Miller,"Gardens as Political Discourse").

LANDSCAPE

The idea of the garden as a landscape is,in the early 2000s,most familiar as the natural landscape garden,or *jardin anglos-chinois,* an artistic bequest of the eighteenth-century British.

The garden as natural landscape rejecting artificiality and the symmetric knots of formal gardens is an extrapolation of Eden from *Paradise Lost* (1667).According to Haorace Walpole,Milton is responsible for popularizing in garden design the idea that "'only after the Fall did man have to invoke art to shore a damaged nature'" (Hunt and Willis,p.79).

During the eighteenth century this account inspired new garden design in England.Since formal gardens exemplified monarchies,the specifics of Milton's description of Eden as a natural landscape rather than as a geometric formal garden (favored by kings) are a function of his intense interest in Puritan antimonarchical politics,adding to their persuasive force.

In 1692 William Temple introduced the Chinese term *sharawadgi,* of which the origin is undertermined (although he thought Chinese),to refer to beauty that imitated nature rather than relying on geometric pruning and symmetrical designes.

Temple defined *sharawadgi* as that sort of oder "where Beauty shall be great,and strike the Eye,but without any Order or Disposition of Parts,that shall be commonly or easily observ'd" (Hunt and Willis,p.98).For half a century Chinese "irregular" gardening principles,known also from Matteo Ripa's illustrations of Chinese gardens,were associated with anti-monarchical (even Whig) politics,until Walpole reversed that association by comparing the Chinese gardens described by the French missionary to China Pere Attiret to French formal monarchical gardens. In addition,visual appropriation of adjacent landscape was integral to gardens in Renaissance Italy,eighteenth-century Britain,and Muromachi,Momoyama,and Edo,Japan,where the term*shakkei,* or "borrowed scenery," was coined to describe it (Nitschke).In Japan,gardens have been imitating nature for nearly one thousand years,and the rules for such gardens were transmitted both orally and in writing (Slawson).

GARDEN AS PICTURE

One variant of the garden as a landscape is the garden based on landscape painting (*Ut pictura hortus*).Christopher Hussey's landmark study *The Picturesque* demonstrated the power of the garden,once it was modeled on painting,to make the "picturesque" a category that could be applied to all landscape—the principle upon which highway scenic overlooks are based.According to Walpole,the early English landscape gardens by William Kent were also designed based upon pictorial compositions.The idea of modeling a garden on a landscape painting has a lively history in East Asia as well,where it can be seen in the dry rock gardens of Daitoku-ji Temple in Kyoto,based on Song Chinese landscape paintings.

COMMUNITY GARDENING

Fig . Strathcona Heights Community Garden in Ottawa,Canada.

Fig. Mobility Community gardening

PURPOSE

Community gardens provide fresh produce and plants as well as satisfying labor,neighborhood improvement,sense of community and connection to the environment.They are publicly functioning in terms of ownership,access,and management,as well as typically owned in trust by local governments or not for profit associations.

Community gardens vary widely throughout the world.In North America,community gardens range from familiar "victory garden" areas where people grow small plots of vegetables,to large "greening" projects to preserve natural areas,to tiny street beautification planters on urban street corners.Some grow only flowers,others are nurtured communally and their bounty shared.There are even non-profits in many major cities that offer assistance to low-income families,children groups,and community organizations by helping them develop and grow their own gardens.In the UK and the rest of Europe,closely related "allotment gardens" can have dozens of plots,each measuring hundreds of square meters and rented by the same family for generations.In the developing world,commonly held land for small gardens is a familiar part of the landscape,even in urban areas,where they may function as market gardens.They also practice crop rotations with versatile plants such as peanuts,tomatoes and much more.

Fig .Crops at the former South Central Farm inLos Angeles,California

Community gardens may help alleviate one effect of climate change,which is expected to cause a global decline in agricultural output,making fresh produce increasingly unaffordable.Community gardens encourage an urban community's food security,allowing citizens to grow their own food or for others to donate what they have grown.Advocates say locally grown food decreases a community's reliance on fossil fuels for transport of food from large agricultural areas and reduces a society's overall use of fossil fuels to drive in agricultural machinery.A 2012 op-ed by community garden advocate Les Kishler examines how community gardening can reinforce the so-called “positive” ideas and activities of the Occupy movement.

Community gardens improve users’ health through increased fresh vegetable consumption and providing a venue for exercise.The gardens also combat two forms of alienation that plague modern urban life,by bringing urban gardeners closer in touch with the source of their food,and by breaking down isolation by creating a social community.Community gardens provide other social benefits,such as the sharing of food production knowledge with the wider community and safer living spaces.Active communities experience less crime and vandalism.

Ownership

Land for a community garden can be publicly or privately held.One strong tradition in American community gardening in urban areas is cleaning up abandoned vacant lots and turning them into productive gardens. Alternatively, community gardens can be seen as a health or recreational amenity and included in public parks,similar to ball fields or playgrounds. Historically,community gardens have also served to provide food during wartime or periods of economic depression. Access to land and security of land tenure remains a major challenge for community gardeners and their supporters throughout the world,since in most cases the gardeners themselves do not own or control the land directly.

Some gardens are grown collectively,with everyone working together; others are split into clearly divided plots,each managed by a different gardener (or group or family).Many community gardens have both “common areas” with shared upkeep and individual/family plots.Though communal areas are successful in some cases,in others there is atragedy of the commons,which results in uneven workload on participants,and sometimes demoralization,neglect,and abandonment of the communal model.Some relate this to the largely unsuccessful history of collective farming. Unlike public parks,whether community gardens are open to the general public is dependent upon the lease agreements with the management body of the park and the community garden membership. Open or closed-gate policies vary from garden to garden. However,in a key difference,community gardens are managed and maintained with the active participation of the gardeners themselves,rather than tended only by a professional staff. A second difference is food production: Unlike parks,where plantings are ornamental (or more recently ecological),community gardens often encourage food production by providing gardeners a place to grow vegetables and other crops.To facilitate this,a community garden may be divided into individual plots or tended in a communal fashion,depending on the size and quality of a garden and the members involved.

Plot size

In Britain,the 1922 Allotment act specifies "an allotment not exceeding 40 [square] poles in extent"; since a rod,pole or perch is 5.5 yards in length,40 square rods is 1210 square yards or 10890 square feet (equivalent to a large plot of 90ft x 121ft).In practice,plot sizes vary; Lewisham offers plots with an "average size" of "125 metres square". In America,plots vary; for example,plots of 3m × 6m (10ft × 20ft = 200 square feet) and 3m x 4.5m (10ft x 15ft) are listed in Alaska.Montgomery Parks in Maryland lists plots of 200,300,400 and 625 square feet.In Canada,plots of 20x20 and 10x10 feet,as well as smaller "raised beds",are listed in Vancouver.

Plant choice and physical layout

While food production is central to many community and allotment gardens,not all have vegetables as a main focus.Restoration of natural areas and native plant gardens are also popular,as are "art" gardens.Many gardens have several different planting elements,and combine plots with such projects as small orchards,herbs and butterfly gardens.Individual plots can become "virtual" backyards,each highly diverse,creating a "quilt" of flowers,vegetables and folk art.

Group and leadership selection

The community gardening movement in North American prides itself on being inclusive,diverse,pro-democracy,and supportive of community involvement.Gardeners may be of any cultural background,young or old,new gardeners or seasoned growers,rich or poor.A garden may have only a few people active,or hundreds. Finally,all community gardens have a structure.The organization depends in part on whether the garden is "top down" or "grassroots".There are many different organizational models in use for community gardens.Some elect boards in a democratic fashion,while others can be run by appointed officials.Some are managed by non-profit organizations,such as a community gardening association,a community association,a church,or other land-owner; others by a city's recreation or parks department,a school or University. Gardeners may form a grassroots group to initiate the garden,such as the Green Guerrillas of New York City,or a garden may be organized "top down" by a municipal agency.The Los Gatos,California-based non-profit Community Gardens as Appleseeds offers free assistance in starting up new community gardens around the world.

Membership rules and fees

In most cases,gardeners are expected to pay annual dues to help with garden upkeep,and the organization must manage these fees.The tasks in a community garden are many,including upkeep,mulching paths,recruiting members,and fund raising.Rules and an 'operations manual' are both invaluable tools,and ideas for both are available at the ACGA.

EXAMPLES

Australia

The first Australian community garden was established in 1977 in Nunawading,Victoria followed soon after by Ringwood Community Garden in March 1980.

Spain

Most older Spaniards grew up in the countryside and moved to the city to find work.Strong family ties often keep them from retiring to the countryside,and so urban community gardens are in great demand.Potlucksand paellas are common,as well as regular meetings to manage the affairs of the garden.

United Kingdom

In the United Kingdom,community gardening is generally distinct from allotment gardening,though the distinction is sometimes blurred.Allotments are generally plots of land rented to individuals for their cultivation by local authorities or other public bodies—the upkeep of the land is usually the responsibility of the individual plot owners.Allotments tend (but not invariably) to be situated around the outskirts of built-up areas.Use of allotment areas as open space or play areas is generally discouraged.However,there are an increasing number of community-managed allotments,which may include allotment plots and a community garden area,many of them overseen by the Federation of City Farms and Community Gardens (a registered charity). The community garden movement is of more recent provenance than allotment gardening,with many such gardens built on patches of derelict land,waste ground or land owned by the local authority or a private landlord that is not being used for any purpose.A community garden in the United Kingdom tends to be situated in a built-up area and is typically run by people from the local community as an independent,non-profit organisation (though this may be wholly or partly funded by public money). It is also likely to perform a dual function as an open space or play area (in which role it may also be known as a 'city park') and—while it may offer plots to individual cultivators—the organisation that administers the garden will normally have a great deal of the responsibility for its planting,landscaping and upkeep.An example inner-city garden of this sort is Islington's Culpeper Community Garden,which is a registered charity,or Camden's Phoenix Garden.

Taiwan

There is an extensive network of community gardens and collective urban farms in Taipei City often occupying areas of the city that are waiting for development.Flood-prone river banks and other areas unsuitable for urban construction often become legal or illegal community gardens.The network of the community gardens of Taipei are referred to as *Taipei organic acupuncture* of the industrial city.

Mali

Often externally supported,community gardens become increasingly important in developing countries,such as West African Mali to bridge the gap between supply and requirements for micro-nutrients and at the same time strengthen an inclusive development.

2

Home Garden Technology

IMPROVE FAMILY NUTRITION BY DEVELOPING YOUR HOME GARDEN

WHAT IS THE PURPOSE OF THESE LEAFLETS?

The purpose of the leaflets is to help farmers improve family food supplies and nutrition year round through home gardening.

This package contains 15 technology leaflets with ideas and technical recommendations on how to improve family food supplies and nutrition through home gardening.

Each leaflet provides information on a technology option or on the type of improvements farmers can make in their home garden to increase food production,to provide a greater diversity of plant foods and to add nutritional value to the family's daily diet.

Who are the leaflets for?

The leaflets are intended for use by agricultural extension workers and farmers who are able to read.The leaflets should be used in situations where a farm family wishes to:

- set up a new home garden for family food production and income;
- develop or expand an existing home garden to improve food production and diversify crops;
- improve family food supplies and nutrition.

How should the leaflets be used?

The leaflets provide basic information,ideas and suggestions on different technology options or home garden improvements.They can be used either singly or in combination with one another,depending on the type of improvements farmers wish to make.Agricultural extension workers should assist farmers in selecting the technology they want to adopt in accordance with the kind,variety and quantity of home garden crops they want to grow.

Farm families should always contact their agricultural extension worker if they need help or advice with technical farming issues such as crop management,pesticide use,water management and many other topics.

HOME GARDEN TECHNOLOGY: THE HOME GARDEN

WHAT IS A HOME GARDEN?

This leaflet aims to help you understand the many different things that make up a typical home garden.When you know how the home garden is made up (its structure) and what it does and produces (its functions or outputs) you will be able to improve the home garden's output to suit your own family's needs.

The home garden is an integrated system which comprises different things in its small area: the family house,a living/playing area,a kitchen garden,a mixed garden,a fish pond,stores,an animal house and,of course,people.It produces a variety of foods and agricultural products,including staple crops,vegetables,fruits,medicinal plants,livestock and fish both for home consumption or use and for income.

WHAT ARE THE BASIC STRUCTURE AND FUNCTIONS OF THE HOME GARDEN?

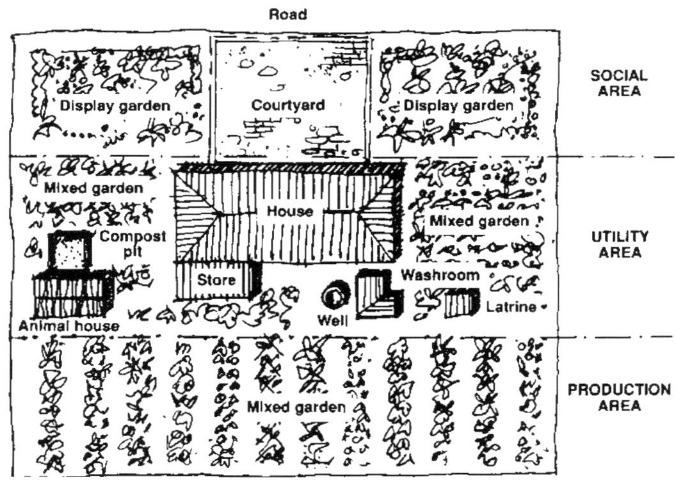

Fig. Basic home garden map

The social area

Location: in front of the house,incorporating the clean-swept courtyard.

Use: mostly a place for social activities - meeting and talking,children's play,display gardens and also for drying grain. The home garden compound has different areas and functions.There are three main areas within the typical home garden.Each of these provides different things for the family that lives there.The areas divided only by dotted lines because,in reality,the functions of the areas overlap the imaginary dotted boundaries.

The utility area

Location: around the house.

Use: mostly a place for physical objects or activities - living, washing, storage, animal house and latrine but also kitchen garden.

The production area Location: the rear part of the garden.

Use: mostly a place for growing food and cash crops and raising animals (e.g.fish,chickens,pigs).

WHAT IS THE SIGNIFICANCE OF THE HOME GARDEN AMONG THE FAMILY'S FARMLANDS?

Most families have more than one area of land for farming.Usually a family has a home garden area and a food cropping area near to the village.Together,these areas of land make up the family's farmlands.The family divides its working time and resources between these two areas.Each area of land is used in a different way but,together,the two areas must provide all the family's needs.

The home garden has a special significance because among a family's most important basic needs are food and shelter.If developed well,the home garden can provide:

- *Enough nutritious non-staple foods for all the family year round,* including extra food stocks for processing and sale to obtain income and a reserve for special occasions or emergencies (e.g.sometimes a staple food crop is lost in a flood,eaten by pests or reduced because the farmer falls sick and cannot work for some time).
- *Income from the sale of home garden produce.* Sales of home garden produce can contribute considerably to a family's income (to buy daily essentials and farming inputs that cannot be produced on the family's farmlands as well as other goods and services).
- *Farm development.* The home garden has a plant nursery for growing plantation seedlings,for trying out new farming ideas and crops and for processing and storing seeds for the next planting season.

HOW TO APPRAISE YOUR HOME GARDEN

Before trying to improve your home garden,you have to find out more about it; particularly why it is not producing more food or income or providing inputs for farm development.There are many things to find out because the home garden has different functions,i.e.social,utility and economic functions.You should allow at least one hour for the appraisal.

Step 1: Get the right people to participate

Different people know different things about home gardens.The farmer and the housekeeper are the most important because they know the home garden's history and what the home garden provides for their family.

The local agricultural extension agent will be able to help identify plants and to assess the soil and other technical aspects.You may want to ask other people to participate,for example your neighbours,relatives or women's farmer group members.

Step 2: Make a map of the home garden

Make a map of your home garden with the help of the others.One way of doing this is by drawing a "mud map" on the ground with a stick and using stones,leaves and other materials to represent the locations of major features such as trees and areas for food crops,vegetables,herbs,buildings and activities.Mark areas where the land is sloping or swampy.

Step 3: Make a copy and keep it

Copy the map as clearly as possible on to paper with your notes.The map will make it much easier to think about the chances and improvements you want to make.

Fig. Discussing and drawing the home garden map

Home garden technology: Planning improvements to the home garden

Every home garden can be improved to fulfill your family's needs better.A well-planned and well-tended home garden can provide nutritious food,income,medicines,seeds and seedlings for the family's other land areas.At the same time,it will be a beautiful place to live in.To improve your home garden,you need to know four things:

- what your home garden produces now;
- what you would like your home garden to produce in future;
- how you can improve your home garden;
- what inputs are needed.

STEP 1: KNOW YOUR HOME GARDEN

Make sure you have a good idea of the structure and main functions of your home garden.A map of your home garden will help you and others assisting you to visualize exactly what your home garden is and what it can do.

STEP 2: SET OBJECTIVES

Make a list of the main things you want your home garden to produce.You should also identify the major constraints you need to overcome,such as wild pigs,poor soil or sloping land.Some examples of objectives.

Planning improvements and changes to your home garden requires some thinking.Deciding what you want to do depends on your situation,such as soil type,sloping land,and how much time and money you have available.Your wife or husband and maybe a friend should participate in the planning.Also ask your agricultural extension agent for technical advice to help you make the right decisions.

STEP 3: SELECT TECHNOLOGY OPTIONS

Choose those technology options that meet your needs and situation.Note that you can choose either a single technology option or a combination of several options,depending on the type of land and resources that you have.

Using the home garden map,identify where the technology options should be located.The technology options sometimes overlap,for example Living fences are useful around an Intensive Vegetable Square.Walk around the home garden with the map and try to imagine how the technology options you have selected will fit together into a system.

Table. Technology Options

Leaflet number	Technology option	Leaflet number	Technology option
3	Growing plants for daily nutrition nutrition	10	Living fences
4	Planting crops for a continuous food supply	11	Multiple cropping
5	Soil improvement	12	Intensive vegetable square
6	Use of sloping land	13	Multilayer cropping
7	Cover cropping	14	Growing fruit- and nut- trees
8	Using wetland	15	Home garden nursery
9	Safe and effective crop protection		

Each of these technology options are briefly described in the respective leaflets,and you can probably recognize many of them in the home gardens and farmed land of your village. Use the map as a vision of the garden you want to create.Everyone who works in the garden,including advisers,should refer to the map.In this way it is easier to ensure that each step in the completion of the garden is thought about,discussed and understood.

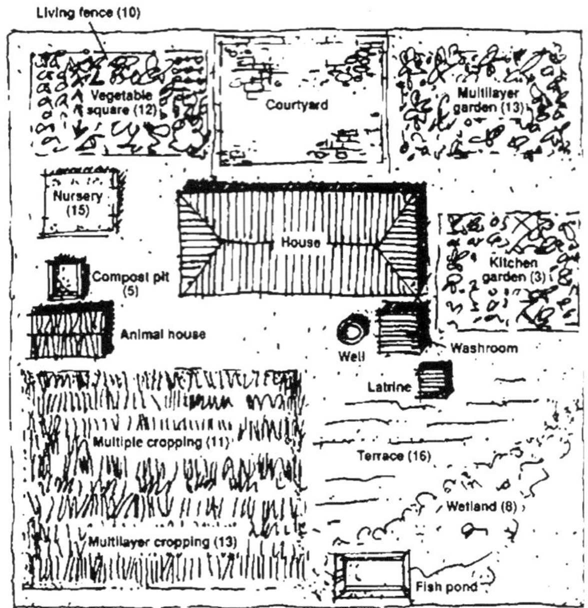

Fig. Example of home garden map with different technology options

Home garden technology: Growing plants for daily nutrition

Nutrition is about all the aspects of food and how it is used in the body.Most people eat because they are hungry.However,while the feeling of hunger tells you to eat,it does not tell you what to eat.This leaflet gives a brief description of some of the main nutrients that make up food,why each nutrient is needed and which kinds of home garden foods provide healthy meals.

FOOD IS MADE UP OF A COMBINATION OF NUTRIENTS

Food is made up of nutrients such as carbohydrates,fats,protein and micronutrients (vitamins and minerals).Nutrients are needed for energy (working and playing),for growth (building and maintaining the body) and for protection against infection. Many foods contain several nutrients; for example,rice,groundnuts and soybeans contain carbohydrates and fats for energy,protein for body building and small amounts of vitamins and minerals for protection.Green leafy vegetables such as pumpkin leaves and orange fruits are very rich sources of vitamins A and C for protection.Animal foods such as fish,chicken and eggs are also rich sources of nutrients,especially protein and carbohydrates,and some vitamins and minerals.

NUTRIENTS ARE NEEDED TO KEEP THE BODY HEALTHY

Plants require certain types and quantities of nutrients in the course of their life to keep them alive and healthy.In the same way,people need a sufficient variety of nutrients from conception to old age.Small children and pregnant or lactating mothers,especially,must have enough nutritious food to ensure proper growth,mental development and health.

It is essential to eat a variety of plant foods every day in order to remain healthy and well-nourished.Also,animal foods such as fish,chicken and eggs should be eaten as often as they are available.

MAKE A KITCHEN GARDEN TO PROVIDE A VARIETY OF FOODS

The easiest way to get a variety of nutritious foods on a daily basis is from a kitchen garden.Located near the kitchen,the garden can be watered and fertilized with kitchen wastes very easily.When a mother is preparing a meal,she need only take a few steps outside to pick herbs,green leaves,spices,vegetables or fruits from the garden.

EAT A VARIETY OF NUTRITIOUS FOODS EVERY DAY

Table. Nutrients from home garden foods

Energy	Protein	Fat	Vitamin A	Vitamin C
Avocado	Cashew nut	Avocado	Fruit	Cashew fruit
Banana	Cowpea	Cashew nut	Banana	
Breadfruit	Eggs	Coconut milk	Bitter cucumber	Custard apple
Canna root	Fish	Coconut oil	Canistel	Guava
Cashew nut	Groundnut	Groundnut	Mango (ripe)	Litchi
Cassava	Koro bean	Milk	Papaya (ripe)	Longan
Coconut flesh	Long bean	Butter (ghee, etc.)	Pumpkin	Mango
Coconut oil	Meat			Papaya (ripe)
Groundnut	Milk		Leaves	Pineapple
Jackfruit	Mung bean		Amaranth	Rambutan
Maize	Pigeon pea		Bitter cucumber	Soursop
Rice	Sesbania grandiflora		Cassava	Tomato
Sugar cane	Soybean		Drumstick tree	
Sweet potato	Wing bean		Gnetum gnemon	
Taro root			Papaya	
Yam			Pumpkin	
			Katuk (Sauropus sp.)	
			String bean	
			Sweet potato	
			Taro	
			Water spinach	

Green leaves,vegetables or orange and yellow fruit and vegetables should be eaten every day.Most fruit and vegetables taste better and are more nutritious when they are fresh.Children especially like ripe fruits.These taste better because they are full of sugars and vitamins.

Beware,however,because soft,ripe fruits (e.g.papaya) and tender green leaves are easily damaged on a trip to or from the market.Handle and wrap produce carefully and store it in a cool place.

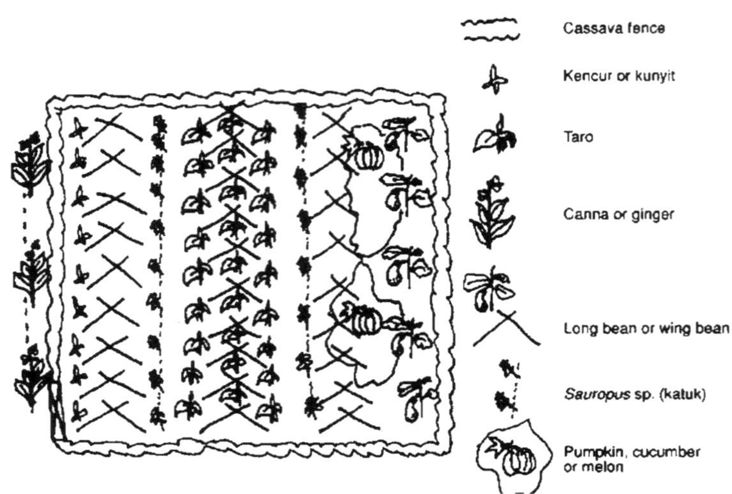

Fig. A kitchen garden

PREPARE A NUTRITIOUS MEAL FOR ALL THE FAMILY

Start with a common starchy staple food (e.g.rice) and combine it with one or more foods from each of the various food groups. The guide shows how you add foods to the staple.Try to add one or more foods from each part of the guide.From the left: legumes and/or food from animals; from the right: fruit and/or vegetables; from above: some energy-rich foods; from below: some flavouring foods.

HOME GARDEN TECHNOLOGY: PLANTING CROPS FOR A CONTINUOUS FOOD SUPPLY

YEAR-ROUND FOOD SUPPLY

Every family should have access to enough nutritious food to ensure that all its members stay active and healthy.Food can be produced on the family's land or bought with money from sales of crops,or earned from an off-farm job.However,the best security is a home garden which always produces food for home consumption year round.

SAVE CASH BY GROWING YOUR FOOD

Many villages are located at some distance from towns and markets.Food supplies coming from outside are often expensive and difficult to transport,especially if heavy rain has flooded or damaged roads.A well-developed home garden can supply sufficient food for consumption on a daily basis.Growing your food at home saves money and effort and ensures a regular supply of food when roads are cut off.

SELL EXCESS FOR CASH

Off-farm employment can provide cash income.However,it is not a reliable source of income.Activities such as road construction and tree felling are available for a limited period of time only.Your home garden can provide cash from sales of crops such as fruit,vegetables and processed foods made from soybean,cassava and coconut year round.

AVOID TOTAL CROP LOSS THROUGH PLANT DIVERSITY

Farm crops are sometimes ruined by wild animals,drought,flood,pests or diseases,especially if they are in a monocropping system.Crop diversity in the home garden reduces the spread of plant diseases and ensures that many food plants survive even if there is a flood or drought.Root crops,fruit- and nut-trees should be interplanted with staple food crops,legumes and vegetables.Crops that take longer to mature are mixed with shorter-maturing plants in the multiple cropping method of farming.Fill the home garden by mixing plants of different heights for multilayer cropping,and use all available areas,even swamp or slopes.

HOME GARDEN FOOD RESERVES

It is very important to have reserves of food or money (or both) so that your family can live through emergencies or special occasions.For example,if you fall sick and cannot work or your staple food crop is ruined somehow,you may need to raise some money quickly.

Table. Suggested minimum target plantings for the home garden

Plants for moist areas	**Plants that cover the soil**
Taro	Gourd
Water spinach	Vine legumes
Sugar cane	Sweet potato
Banana	
Lemon grass	
Plants to grow on a trellis	**Plants as living fences**
Wing bean	Leucaena
Koro bean	Drumstick tree
Long bean	Sesbania sp.
Pumpkin	Cassava
Gourd	Pineapple
Passionfruit	Lemon grass
Yam	Gliricidia sp.
Pepper	
Bitter cucumber	
Plants to grow under a trellis (shade)	
Most leafy plants	
Some root crops, e.g. taro, sweet potato	

Fruit-trees can give a continuous supply of food throughout the year or in different seasons.Find out the harvest times of different fruits in your area and plant in such a way as to have fruit all year.

Root crops are living food stores which can be left in the ground until you need them.Many also provide nutritious leaves (e.g.cassava,sweet potato,amaranth).

Chickens and other animals can be fed on household scraps and home garden plants.You can keep them for sale or for eating when you need to.

Crop	Minimum number	Frequency / Every three months
Root crops		
Sweet potato	100	+
Taro	150	+
Yam	50	+
Cassava	300	+
Legumes		
Groundnut	600	+
Soybean	600	+
Mung bean	600	+
Long bean or French bean		
Vegetable crops		
50		+ (upland)
Water spinach		+
Pumpkin	4	+
Amaranth	25	+
Rape or jute	50	
Katuk (Sauropus sp.)	100	
Cassava leaves	100	
Fruit		
Papaya	5	
Coconut	1 5	
Banana	1 5	
Jacktruit	5	
Guava	5	
Citrus	5	
Spices and medicinal plants		
Lemon grass	5	
Chili	5	+
Slack pepper	10	+
Garlic or onion	20	+
Ginger	10	+

Table. Suggested crop locations In the home garden

Home garden technology: Soil improvement

WHAT IS IMPORTANT IN SOIL?

Good soil is essential for a good harvest.Soil must have all the nutrients necessary for plant growth,and a structure that keeps plants firm and upright. The soil structure must hold enough air and water for plant roots,but must allow excess water to drain away.

THE LIVING SOIL SYSTEM

Most nutrients are naturally recycled from the soil through plant roots and back to the soil through fallen leaves and other organic matter.Worms,insects and tiny organisms such as fungi feed on organic matter and change it into humus which makes topsoil dark and gives it a good structure.Humus is quickly lost or washed away if the soil is left exposed.Subsoil is usually less fertile.

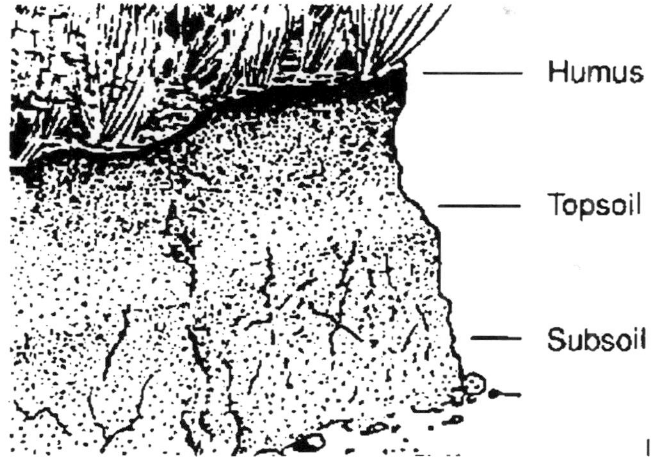

Fig. Topsoil is the best part

THERE ARE DIFFERENT TYPES OF SOIL

Some soil is naturally fertile (such as river plain soil or volcanic soil) but,in many places,the soil is naturally infertile or has lost nutrients through clearing,regular burning or continuous cropping without the application of fertilizer.

Table. Common soil types and treatments

Soil type	Features	Methods of improvement
Sand	- Poor structure	- Regularly add organic matter and fertilizer, use green manure crops
	- Poor fertility	
	- Cannot hold water	
Silt	- Poor structure	- Add coarse organic matter
	- Good fertility	
Clay	- Dries hard	- Add organic matter, compost and gypsum
	- Holds too much water	
Acid subsoil	- Subsoil layer is toxic to some plants	- Keep soil inundated (rice paddy)
		- Grow shallow-rooted plants (vegetables)
		- Apply ground limestone (3 kg/10 m^2)

PLANT NUTRITION

Good crops will only grow if there are enough nutrients in the soil.

Table. Nutrients and their functions

Nutrient	Function	Deficiency symptoms	Sources
Nitrogen (N)	Growth in leaves and stems	Pale green or yellow leaves	Urea, ammonium nitrate, ammonium phosphate (MAP or DHAP), NPK or other nitrate fertilizer
	Green colour and pest/disease resistance	Poor growth	
		Leaf fall	Animal waste
		Pest problems	Compost
			Green manure crops
Phosphorus (P)	Beans, seeds and fruit(early maturity)	Stunted growth	Superphosphate, MAP, DHAP, NPK
	Root formation	Diseases	Chichen manure
	Drought resistance	Poor formation of side shoots and flowers	Ash Ground animal bones
Potassium (K)	Strong roots and stems	Curled, wrinkled or burnt leaves	Potassium chloride (muriate of potash), potassium nitrate, NPK
	Fat seeds and fruits	Uneven ripening	Ash
	Helps move nutrients around the plant	Poor growth	
			Manure
			Banana leaves and stems
			Maize cobs
			Compost

HOW TO MANAGE SOIL FERTILITY

Some chemical nutrients in the soil are stable (e.g.phosphorus) while others are quickly lost or consumed (e.g.nitrogen).

A farmer needs to make a basic application of enough nutrients to start a garden and then maintain the supply of nutrients by regular applications as the crops grow.Poor soil will become productive if properly managed.Manure and compost are needed to improve soil structure while chemical fertilizer is needed for a higher production.

The general method is to dig compost,organic matter,manure or chemical fertilizer into the soil just prior to planting.This is the basic application.After planting,apply small amounts of manure,compost or chemical fertilizer alongside plants about every two weeks until harvest.

Fertilizer

The quickest way to put plant foods into the soil is to use chemical or mineral fertilizers containing one or more of the three chemical nutrients needed by plants. Fertilizers can wash away quickly so do not apply them too early before planting.Fertilizers cost money and are very concentrated so you only need to apply about one handful for every 4 m^2.Never put fertilizer in a heap too close to a plant or it may burn the roots or stem of the plant.It is better to spread the fertilizer out and lightly mix it into the soil surface.

Compost

Compost is easy to make and does not cost anything if you have the time,some space in your garden and access to materials such as animal and kitchen waste,leaves and grasses. in the ground below the pit.

Fig. Example of a compost heap

Fig. Hedgerow use

Compost pits are common but they take time to dig and nutrients are lost It is better to make a compost heap.Make the compost in layers and add kitchen refuse every day.Turn or mix the compost heap every month to help it rot and break down.It takes three to four months to become dark and ready to use.Keep the heap in place with logs,banana stems or bricks around the edge.

Green manure and compost crops

Another way to feed the soil is to grow green manure crops and dig them into the soil after cutting.These crops are also very good for compost,especially legume plants (such as leucaena,Flemingia sp.pigeon pea and centre) which collect nitrogen.Legume trees such as leucaena can be grown above or near the food crops and their branches occasionally pruned off and left on the ground as manure.Low legume plants can be planted with a food crop to help improve the soil and keep out weeds.

Table. Green manure crops

Hedgerow green manure crops	Green manure/compost cover crops
Leucaena	Grasses
Flemingia sp.	Centro (Centrosema sp.)
Gliricidia sp.	Puero (Pueraria sp.)
Pigeon pea (Cajanus sp.)	Water hyacinth (swamp)
Guinea grass	
Setaria sp.	

Using compost and Manure

Manure can be dried in the shade (for example under a stable) and stored for later use.Fresh manure may burn plants if placed too close. Compost is best when it is crumbly like forest litter and not heavy or sticky.Compost and manure can be mixed into the soil in a hole before planting a tree or dug into a garden before planting vegetables or food crops.

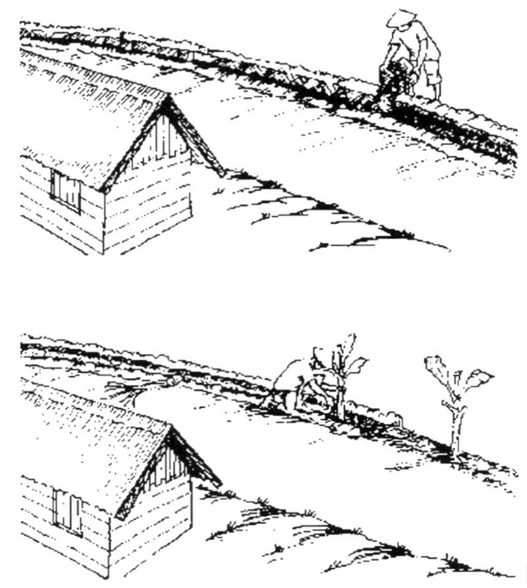

Fig. Compost trenches

A compost trench is useful for a row of new crops or for feeding established crops.Compost and manure can be sprinkled on the soil surface,but it is better if it is protected from full sunlight.Fully rotted compost is good to mix with sandy soil for use in a nursery.

Mulch

Another way to feed the soil is using mulch,which protects soil from erosion and reduces weeds. Mulch materials such as straw or green manure cuttings should be spread about 4 to 6 cm thick around plants.

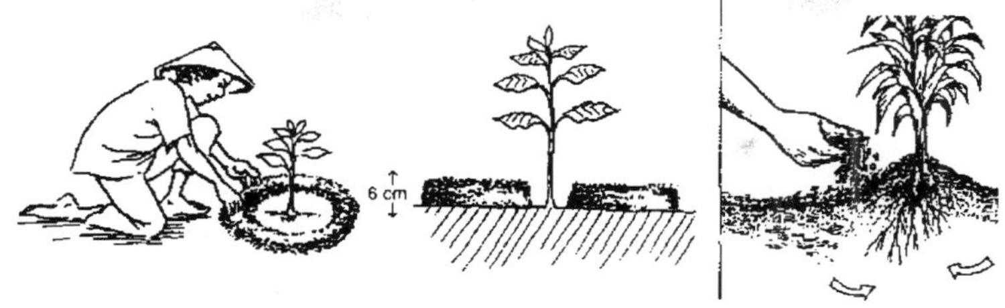

Fig. Mulch

HOME GARDEN TECHNOLOGY: USE OF SLOPING LAND

All of the home garden area can be used to grow useful plants,but sloping land needs special care to keep the soil in good condition.

PREVENT EROSION

The best part of the soil is the dark layer of topsoil,which takes many years to develop.

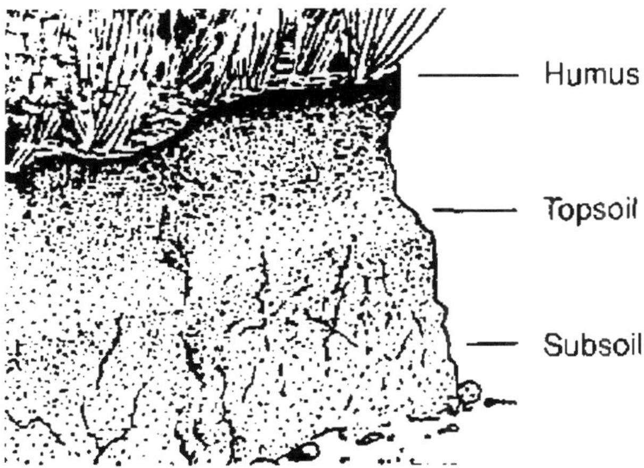

Fig. Topsoil is the best part

Topsoil is rich in plant nutrients and beneficial soil organisms such as worms.Under the topsoil is the yellow or light brown subsoil which may be very acid and is harder for plants to grow in.Humus is a layer of rotting plant debris which feeds the topsoil.These two layers are lost easily through erosion by rain,wind,cultivation,foot traffic and by the ground being swept clean every day.

Cover the soil

Soil can be covered with mulch or with living c over crops.Covering the soil reduces weeds and prevents the soil from washing away when it rains.

A straw mulch or humus layer also prevents soil from sticking to your feet and "walking" out of the garden,taking fertilizer and nutrients with it.

Fig. Cover cropping

Plant hedgerows

Hedgerows planted horizontally across the slope stop rainwater from moving fast over the soil and carrying off the topsoil.Use hedgerows while useful plants for cover crops and hedgerows.

Fig. Hedgerows

Table. Cover crops and hedgerow plants

Cover crops	Hedgerow plants
Food crops	Food hedgerows
Sweet potato	Pineapple
Water spinach	Salak
Pumpkin,cucumber,melon	Lemon grass
Cassava	
Other	Multipurpose trees
Grasses	Pigeon pea (Cajanus sp.)
Puero (Pueraria sp.)	Cailiandra sp.

Centro (Centrosema sp.)
Stylo (Stylosanthes guianensis)
Calopo (Calopogonium mucunoides)
Mucuna sp (lives 4-7 months)
Cowpea

Flemingia sp.
Leucaena
Sesbania grandiflora

Make barriers to catch soil

Logs,banana stems and horizontal channels catch soil when it moves down hill.Paths wear down quickly,and sloping paths should have wooden steps,otherwise steps cut into the soil may be washed away by floods.

Fig. Channels and barriers

Terraces

Terraces can be found in many home gardens.The most important thing is to protect the sloping part of the terrace by planting grasses or hedgerow crops,otherwise the terraces will slide downwards with erosion.When making a terrace,keep it flat by using an A-frame.

Terraces are an excellent long-term way of increasing the cultivated area of a home garden.

Fig. Terraces

Do not attempt to start making terraces by yourself if you have not had previous experience.It is advisable to ask the agricultural extension workers in your village or a neighbour who has constructed terraces before to show you what to do.

HOME GARDEN TECHNOLOGY: COVER CROPPING

COVER THE SOIL

Covering the soil reduces weeds and prevents the soil from washing away when it rains.Erosion of the humus layer and topsoil from your home garden greatly reduces crop growth and yield at harvest.Soil can be covered with living cover crops or with mulch.Cover cropping is the long-term technique for weed control in the home garden.Mulching is a short-term technique to keep weeds from getting established.

How cover crops work

By covering the soil with trailing vines and leaves,cover crops shade the soil surface and protect it from the impact of falling rain which wears away the soil.A dense mat of cover crop provides strong competition against any weed seed which finds its way into the crop.The competition and the shade make the cover crop the winner over most weeds.For example,cowpea will climb up and smother alang-alang.

Fig. Vegetable cover crops

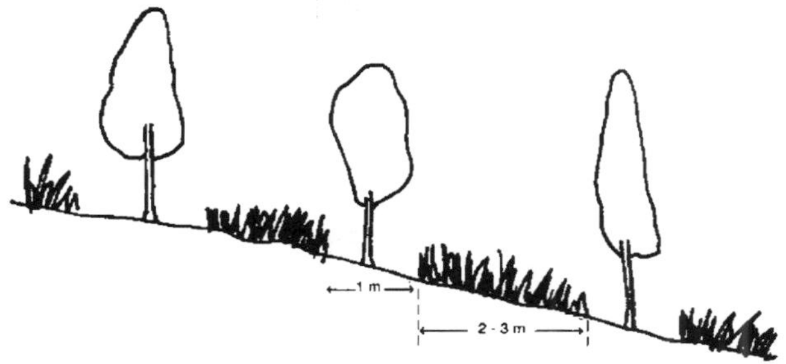

Fig. Cover cropping between tree crops

Two kinds of cover crops

Table. Cover crops

Cover crops	Establishment and care
Food crops	
Sweet potato	- Feed these with compost to help growth of cover and food parts
Water spinach	- Most of these can be interplanted with other food crops
Pumpkin, cucumber, melon	
Other	
Grasses	- Grasses: Uproot a clump and split it into pieces (splits) that include roots and leaves. Plant about 30 cm apart (closer for mall grasses like Manila grass)
Komak	
Puero (Pueraria sp.)	- Broadcast legume seeds or dig them 2 cm into the soil 30-50 cm apart. Hard, dry seeds may need to be scarified before planting
Centro (Centrosema sp.)	
Stylo (Stylosanthes guianensis)	
Calopo (Calopogonium mucunoides)	
Mucuna sp. (lives 4-7 months)	
Cowpea	

Food plants can be used as cover crops.Many other plants can act in the same way.For example,taro plants can be planted close together to become a cover,especially in wet or swampy areas.Another method is multiple cropping where different food crops are planted together,covering the soil.

Most other kinds of cover crops (grasses and creeping legumes) are not food crops.Legumes have friendly bacteria in their roots which provide nitrogen nutrient for the soil.Both grasses and legumes can be cut for green manure,although legumes should not be cut very close to the ground.

How to establish a cover crop

For a food cover crop such as sweet potato or pumpkin,you need to cultivate the soil and mix in some rich compost to feed the growth of the crop.Weeding will need to be done once or twice in the first month until the crop is established.

Plant sweet potato and water spinach using 12 to 20 cm vine cuttings (best with roots),placing them upright or angled with 10 cm of the cutting buried in the soil.Water spinach can also be grown from seed planted I cm deep.The top 2 cm of a sweet potato tuber with shoot buds can also be planted in the same way.

Sweet potato cuttings about 30 cm long (with roots) or the tops of the tuber (with shoot buds) are planted about 30 cm apart.It is best to plant the sweet potato in rows and to dig in a compost trench under the row before planting to help the tubers grow.

For pumpkin,plant three or four seeds for each planting hole.Mix in two handfuls of compost per hole and place the seeds 4 cm deep.Pumpkins should be planted about I m apart to make a good cover crop.

Pumpkin,cucumber and melon are grown only from seed,planted 2 cm deep.

Legume cover crops such as cow pea will not usually require weeding and can survive without compost on reasonable soil.Ask your extension agent for advice on where to get planting materials and how to treat seeds.

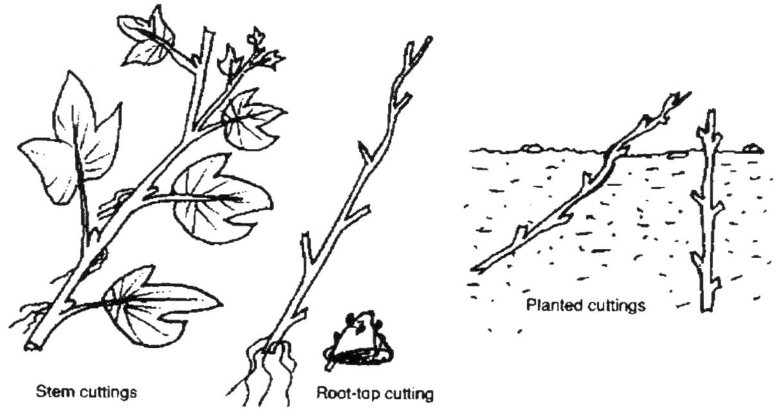

Fig. Planting sweet potato cuttings

HOME GARDEN TECHNOLOGY: USING WETLAND

Low-lying and swampy land can be very productive for many crops,including rice.In a home garden,even a very small area of wetland,such as the banks of a drain,can be used for growing food all year.

THE SURJAN SYSTEM

Management of water is the key to home garden development in low-lying areas prone to flooding.Soils that flood frequently can be made more productive with good water drainage.The home garden using the "surjan" system,which is made by digging soil from wide water channels and piling it up into flattened mounds (beds).In the channels,wet rice,wet taro and water spinach can be grown,sometimes together with fish.On the raised beds,annual food plants and/ or tree crops,especially citrus,can be grown.

Fig. Surjan system

THE SWAMP GARDEN AND FISH POND

Where there is a gully or small creek,a pond can be dug out and allowed to fill with water.If it is necessary to make a small dam,use clayey soil (subsoil) in the dam wall because topsoil and any organic matter,such as sticks and plant stems,will let water through.Also make sure there is an overflow channel which guides water around the dam wall if there is very heavy rain and the pond floods.Without an overflow channel the dam wall may weaken and collapse.

Cover the banks of the pond or channel with useful crops to prevent the soil from washing into the pond and to make use of the land.Depending on the water depth,set some plants in the bottom of the pond.Climbing plants,especially pumpkin and gourds,can be grown on a trellis over the pond to shade the water and keep it cool.

Table. Some useful wetland plants

Plants	Use
Water spinach	Edible stems and leaves
Taro	Edible leaves and root bulbs
Lotus	Edible leaf stems,root bulbs and
seeds	
Banana	Fruit,leaves for packaging
Rice	Grain
Water hyacinth	Compost
Ginger	Edible leaves
Sugar cane	Stems
Lemon grass	Leaves for flavouring

Fish such as tilapia and nila will grow well feeding on tiny organisms in the mud at the bottom,but they must also be fed daily with soft green leaves such as cassava.Growing plants such as water spinach in the pond will provide a place for the fish to hide from the sun or from predators.

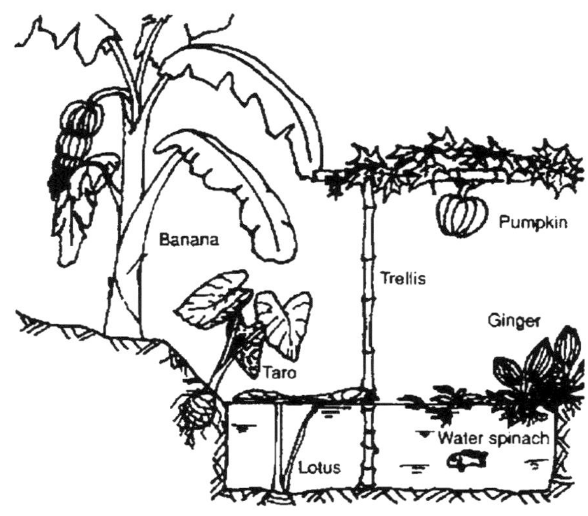

Fig. Fish pond and swamp land

HOME GARDEN TECHNOLOGY: SAFE AND EFFECTIVE CROP PROTECTION

WHAT ARE PESTS AND DISEASES?

A good farmer must know how to manage pests and diseases of crops and,to do this,he or she must understand what pests and diseases are.The first skill to learn is how to recognize what kind of pest or disease is causing the problem.Here are some simple points to remember:

- Pests and diseases are all living things - insects,fungi and bacteria.Generally,they cannot survive for long without a place to live.
- Insects can sometimes be seen on plants or in the soil.They mostly damage plants by chewing holes in roots,leaves and fruit or by sucking sap out of the leaves,stems and fruit.Not all insects are pests,some (e.g.bees) pollinate flowers so that crops have fruit and seeds.
- Fungi are very small but sometimes can be seen as mushrooms.They attack all parts of plants.Signs of fungi may be a powdery substance under leaves,rotten patches,black spots on stems,leaves and fruit or wilting because of rotten roots.Fungi may spread through rain that is splashed up from soil on to plants or they may be carried by wind from one plant to another.
- Bacteria and viruses are only visible with a microscope.They may cause rot in roots and stems,oozing sap,distorted or striped leaves,black spots and other symptoms.They spread through water,soil and affected plants.

Table. Some common pests and diseases in Southeast Asian home gardens

Pests or disease	Plants attacked	Symptoms	Controls
Root rot (Pythium sp. or Phytophthora sp.) (Fungus)	Papaya and many others	Wilting	Plant in well-drained soil
		Collapse of plant	Do not replant where root rot has occurred
		Rotten roots and stem	Avoid introducing root rot in contaminated soil
Bacterial wilt (Pseudomonas solanaceraum)	Ginger, tomato and many others	Wilting and yellowing of leaves	Plant only disease-free seed
		Blackening inside stem	Do not replant with susceptible crops
			Plant a legume crop
			Burn affected plants
Mosaic virus	Papaya	Yellow, stunted leaves	Destroy affected plants
			Plant only disease-free seed
Mealybug and scale insects (various species)	Citrus and many others	Wilting	Wipe insects off plants by hand
		Wax-covered insects on stems	Spray with appropriate pesticide or light oil
			Encourage ladybird insect predators
Tomato grub (Heliothis sp.)	Many vegetables and maize	Holes in fruit	Remove and kill caterpillars
			Use appropriate pesticide

GOOD FARMING PRACTICES PREVENT PROBLEMS

There are important and simple ways to help plants stay healthy and productive.

- Grow plants where soil,water and light conditions suit them.Papaya needs full sunlight,coffee needs shade.Taro likes wet soil but papaya might get root rot and die in the same place.It is important to select the correct plant for each place in your garden.
- Feed your crops and they will feed you.Yellow leaves,poor growth and small fruit are often due to a lack of water or nutrients in the soil.If a plant is sited correctly and no pests or fungi can be seen under the leaves,the application of fertilizer or manure may improve plant growth.
- Minimize competition.Plant crops with just enough space for each one to grow to full size.Weeds can sometimes grow faster than crops and they take soil nutrients necessary for crop growth.Weeds should be removed before planting.Mulch between crops will prevent many weeds from taking root until the crop is well established and covers the soil.
- Protect plants from strong winds,seasonal dry winds or salty winds from the sea.Wind can reduce growth and damage leaves and flowers.Use multipurpose trees as living fences.
- Try to avoid planting large areas of a single crop (monocropping).If one plant gets sick,the disease spreads rapidly throughout the whole crop unless there is another different crop as a barrier.Interplanting can be a useful technique to increase the number of crops in one place.
- One way to keep fungi away from vine plants (such as pumpkin) is to grow them on a trellis.If the soil is very wet or has a lot of clay in it,try growing plants above the soil in baskets or containers filled with good soil and compost.

PHYSICAL CONTROL METHODS

The safest way to avoid pest and disease problems is to practice good garden hygiene.Remove and burn plants affected by diseases before the disease spreads.Dead branches,fallen fruit and tall,dense weeds can house pests and diseases.Remove and burn or compost materials where pests and diseases live and breed.Keep the compost heap away from growing vegetables.Do not replant the same kind of crop in the same place; plant a different kind of crop instead.

NATURAL PESTICIDES AND DETERRENTS

There are many household items which can deter insects.Sucking insects such as aphids can be deterred by sprinkling ash over the insects.They are usually on the underside of leaves.Ash sprinkled around the base of plants can deter some crawling insects.Soapy water poured or sprayed over sucking insects can also be effective.Slugs and other pests can be trapped in a hall: buried bottle with a little beer remaining in the bottom.Coffee grounds will deter many insects. Certain plants are known to repel many types of insect,and some farmers plant these as companions to food crops.Garlic,marigold and lemon grass are some of these plants.

Some farmers know how to prepare natural pesticides from extracts of certain plants,seeds or fruit which can be mixed with water and sprayed on to plants.Some common examples in Southeast Asia are tobacco,neem fruits,rotenone and oil from citrus skin.In general,a farmer must experiment a little to find an effective solution which is easy to prepare.Do not forget that these natural pesticides can also be poisonous to animals and humans.Follow the same safety rules as with chemical pesticides.

SAFE USE OF CHEMICAL PESTICIDES

Occasionally,the use of chemical pesticides is the best method of pest and disease control.Many different pesticides are widely available.They cost money,however,and while they are powerful,they may be ineffective and dangerous if used incorrectly.Using them in a home garden is not the same as using them in a field situation such as a rice paddy.Always read the label on pesticide packages and respect the following rules for safe and effective use of pesticides:

- *Specific targets.*Many pesticides are only effective against specific pests or diseases.Identify the pest or disease causing the problem before you select a pesticide.
- *Protect yourself.*When handling pesticides,especially when mixing or spraying,do not let the chemical touch your body and,if it does,wash it off immediately.Wear gloves or plastic bags over hands,wear a breathing mask or a cloth to filter the air you breathe as well as a hat and a shirt or jacket.
- *Mix correctly.*Follow the instructions on the label to mix the chemical to the right concentration.Do not add more than the rate specified because this may reduce its effectiveness.Repeat applications only according to the frequency written on the label.
- *Withholding period.*After any pesticide is applied to a crop,a certain period must pass before the crop can be harvested and eaten or sent to market.This is the keep-out period.For some pesticides it is only one day,but for others it may be two weeks.Poisoning may occur if the crop is eaten inside the keep-out period.If children or animals cannot be kept away from the crop,do not use the pesticides.
- *Storage and disposal.*Keep pesticides safely locked away from children to avoid poisoning.If there is pesticide left over in the tank after spraying,pour it out but make sure you do this away from streams and ponds so that it does not poison the water and fish.
- *If in doubt.*Do not use a pesticide if you are not sure of the procedures.

HOME GARDEN TECHNOLOGY: LIVING FENCES

PROTECTION WITH PRODUCTION

Food crops need protection from animals and sometimes from people.The idea behind a living fence is that certain plants make good fences and at the same time produce useful things for people,for livestock and for soil improvement.Some plants for making living fences.Some people make good fences using living plants as well as wood or bamboo poles.

TABLE. Plants for living fences

Plants	Notes
- Gliricidia sp.,Leucaena sp.,Sesbania sp.,drumstick tree Salak,pineapple,pandanus Cassava,katuk (Sauropus sp.)	- Leaves are useful for animal fodder it cut regularly at 1.5 m from ground Plant in double rows: Plant 5 cm apart and strengthen the fence with bamboo strips

Goats usually roam free in the village but they can cause much damage in the home garden.Fences should be erected around vegetable areas and food crops.A living fence of lamtoro (Leucaena sp.) planted close together and bound with strips of bamboo will keep them out at the same time as providing them with fodder.Another kind of fence can be made with sticks of cassava bound with bamboo strips.

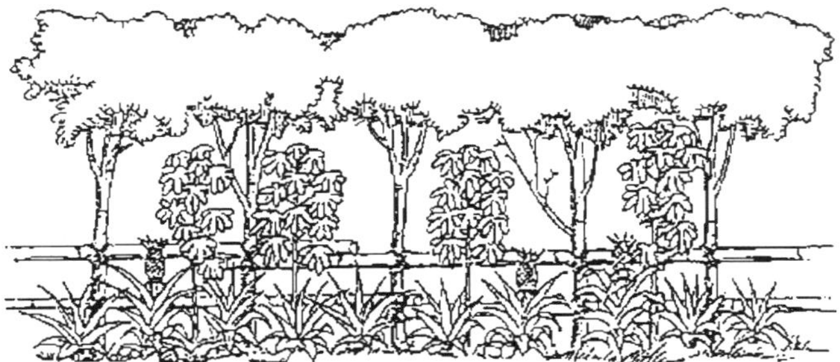

Fig. Living fences with edible leaves

Fig. Salak fence

Wild pigs are smart and they particularly like root crops.One way to deter pigs is with a thick fence of plants that have spines or thorns,such as salak,pineapple or pandanus.

Fig. Pandanus around young coconut

Chickens are usually left free to scavenge for food,including insects and seeds,but they will also strip leaves from plants and seriously damage vegetables and young plants.

Fig. Plant cassava or use bamboo or other sticks around plants

Fig. Protect root crops by placing coconut around the base of the plant

HOME GARDEN TECHNOLOGY: MULTIPLE CROPPING

CONTINUOUS SUPPLY

Overlapping the planting times of several different crops in the same place will provide a year-round supply of food crops and vegetables.In large-scale farming,*monocropping* is common because of the ease of planting and harvesting,but there are problems of weeds and pest attack.Multiple cropping has been practiced for food crops in many countries and is useful in the home garden.

DIVERSITY GIVES HEALTHY PLANTS

Growing different crops together minimizes pest problems and makes efficient use of soil nutrients.Legumes (such as beans) will provide some nitrogen nutrient to other crops such as maize or tomato when planted together.Some plants such as chili,garlic and marigold flowers can keep certain pests away from neighbouring plants.These companion plants and others can be mixed into or around a planted area.

INTERPLANTING AND ROTATION PLANTING

Plants belonging to the same family should not be planted repeatedly in exactly the same place for more than two years,otherwise pests and diseases will build up in the soil.Some of the family groups of plants for rotating around the garden; for example they may be replanted in the next row.It is best to plant legumes before crops of the other families.Cassava,maize and other food crops can be inter planted between other crops.

Table. Vegetables in the same family

	Family	
Solanaceae	**Cucurbitaceae**	**Leguminosae**
Tomato	Cucumber	Groundnut
Chili	Pumpkin	Soybean
Eggplant	Squash	Long bean
Sweet pepper (Capsicum sp.)	Bitter cucumber	Centro (Centrosema sp.)
Sweet potato	Melon	Cowpea
Water spinach		

INTERPLANTING WITH TREES

Tree crops such as coconut,citrus and cinnamon can be planted 6 to 10 m apart.The area in between is good for other crops such as coffee or cocoa,but especially for regular interplanting of annual food crops (e.g.groundnuts,maize,cassava) or vegetables.Monocrop tree crops such as oil palm can be interplanted for the first five to six years,after which they can be underplanted with cover crops.

Fig. Interplanting

FOOD CROP PLANTING SEQUENCES

The sequence of crops planted should follow the changes in season during the year,especially rainy seasons.As a guide for home gardens where hand watering supplements rainfall,crops should be planted in beds or rows according to the example sequences.In the dry season,leaf crops should be planted in the shade and crops such as mung bean and cassava in beds that are watered less frequently.

Table. Examples of crop sequences

Bed						Month						
	8	9	10	11	12	1	2	3	4	5	6	7
1	--Maize-					##### Soybean ###					----Maize---	
	#### Groundnut ##					==== Tomato ===					# Mung bean #	
2	>>>Long bean >>>>					==Sweet potato ===		>>>Long bean>>>			==Sweet potato	
3	+++ Pumpkin +++					### Soybean ###					++ Pumpkin ++	
	:::: Cassava::::											

HOME GARDEN TECHNOLOGY: INTENSIVE VEGETABLE SQUARE

SETTING UP THE VEGETABLE SQUARE

A small area of 30 to 40 m^2 can provide a household with fresh vegetables all year.The idea is to grow different kinds of vegetables one after another on well-fertilized beds.

Step 1

Mark out the square into four planting beds,about I m wide and 5 m long.The beds should be as wide as you can easily manage,leaving some room for paths in between.

Step 2

Cultivate the soil in the beds down to at least 20 cm.Break up the soil with the back of a hoe until it is fine and loose.Mix in about 5 kg of good compost per square metre of bed and add some fine topsoil to raise the level of the bed to about 20 cm above the path.

Step 3

Now you are ready to plant.Make a fence around the vegetable square to keep out wandering animals before any seedlings come up.

CHOICE OF CROPS

Choose crops that will provide good daily nutrition and that the family likes to eat.Remember that tall and ground-level plants can be grown together in a multilayer system,such as long bean on poles above cucumber,or eggplant above sweet potato.Plan the schedule of planting according to the growing time for each type of vegetable.

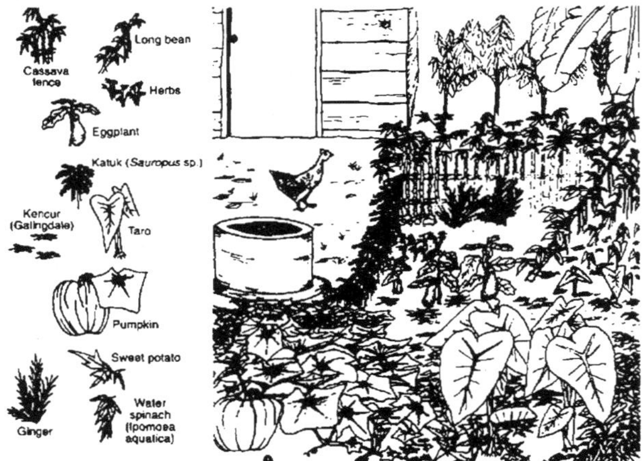

Table. Planting details for selected vegetable crops

Crop	Depth for seed or cutting (mm)	Spacing: Between plants (cm)	Spacing: Between rows (cm)	Harvest (days after panting)
Amaranth	5	8	25	25 60
Long bean	25	50	100'	From 70
Tomato	10	60	45	From 100
Eggplant	10	45	75	From 100
Pumpkin	20	150	150	100
Cucumber	20	200	150	80
Mustard**	5	15	15	40
Celery	5	15	20	25-45
Sweet potato	100	10	75	20 (leaves) 100 (roots)
Water spinach	20	15	30	From 45
Okra	15	45	60	60-90

Plant vegetables that can be harvested together in the same

place.Alternatively,plant fruit vegetables (e.g.tomato, eggplant, maize, chili, cucumber) with leaf and root vegetables (such as water spinach, amaranth, taro, cassava) so that the fruit vegetables can be harvested without disturbing the leaf and root vegetables.

PLANTING

Small seeds (e.g.tomato,mustard,cabbage and amaranth) should be sprinkled in a row,covered lightly,and the seedlings thinned out after they emerge.Alternatively,they can be germinated in a seed bed or nursery and transplanted as seedlings to the garden bed.

Larger seeds (beans,pumpkin) can be planted directly into the garden bed.Seedlings may require shade from direct sun in the first week if there are no trees around the vegetable square.A coconut frond supported by sticks will provide good shade.After seedlings emerge,the bed should be covered with mulch to protect the soil from becoming too hot and drying out the plants.Mulch will also reduce weeds.

Table. Vegetables in the same family

Family		
Solanaceae	**Cucurbitaceae**	**Legurninosse**
Tomato	Cucumber	Groundnut
Chili	Pumpkin	Soybean
Eggplant	Squash	Long bean
Sweet pepper (Capsicum sp.)	Bitter cucumber	Centro (Centrosema sp.)
	Melon	

REPLANTING

Plants of the same family should not be planted repeatedly in exactly the same place for more than two years,otherwise pests and diseases will build up in the soil. Some of the main groups of plants that should be planted in another bed after one or two growing seasons.

It is best to plant legumes before crops of the other families because they increase nitrogen nutrients in the soil.

HOME GARDEN TECHNOLOGY: MULTILAYER CROPPING

NATURAL FOREST

The multilayer structure of natural forest is built up over many years.There are tall trees,medium sized trees and shrubs,climbing vines and leafy shade plants in the shade.This layered structure uses all the sun light available for plant growth,thereby reducing weeds,and keeps the soil healthy.

LONG-TERM CROPPING SYSTEM

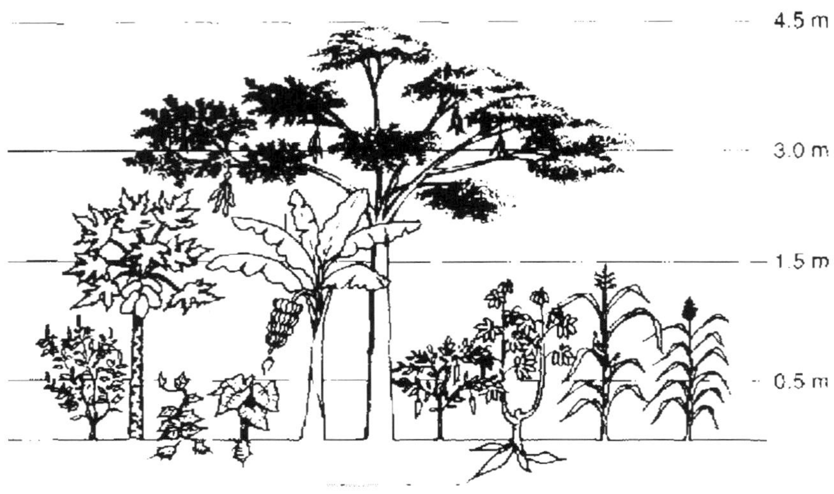

Fig. Heights of plants

In the home garden,the layers can be filled with plants that are of daily use to the household.This system mixes plants with short,medium and long terms before maturity and harvest,similar to multiple cropping.Some ideas for designing parts of the home garden using a system of mixed planting.Plants for different levels.

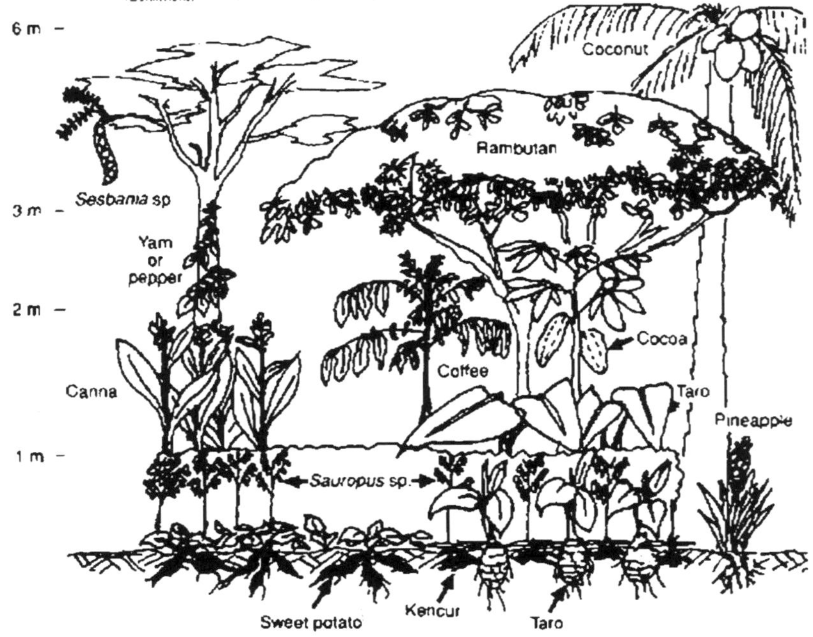

Fig. A multilayer cropping system

Table . Crops for different layers

Crop	Layer
Coconut	Canopy or ceiling
Breadfruit	
Durian	
Jackfruit	Upper middle layer
Jambu	
Rambutan	
Clove	
Cashew nut	
Banana	
Coffee	Lower middle layer
Papaya	
Cocoa	
Soursop	
Pigeon pea	
Taro	Lower layer
Turmeric	
Ginger	
Chili	
Lemon grass	
Sweet potato	Ground layer
Cucumber	
Pumpkin	Climbers
Bitter cucumber	
Yam	
Pepper	
Long bean	
Passionfruit	

TRELLIS

Fig. Trellis systems

HOME GARDEN TECHNOLOGY: GROWING FRUIT- AND NUT-TREES

THE BENEFITS OF FRUIT- AND NUT-TREES

Fruit- and nut-trees are special in the home garden because,unlike vegetables,they will produce for many years.Fruits and nuts are good sources of vitamins,minerals,fats and oils and protein.Fruit is a good snack food for children.Trees are also useful for shade,timber and as a support for climbing plants such as yam,pepper or passionfruit.A selection of different kinds of tree will produce fruit at different times of the year,so the availability of food is spread out.

Where to plant fruit- and nut-trees

All plants grow best where the conditions suit them.Trees occupy the middle and upper layers of the garden and most of them prefer full sunlight.Crops can be planted underneath or between fruit-trees to maximize garden production). Trees can be grown on a range of soil types because they are able to find nutrients and water deep in the soil.Most fruit-trees do not tolerate wetland (banana is an exception).On wetland,dig out canals and use the soil to make raised beds between the canals.Plant fruit-trees (such as citrus) on the raised beds.

Young trees will grow faster if they are sheltered from strong winds or salty winds from the sea.Flowers and fruit can also be knocked off trees such as sugar apple,mango and citrus by too much wind.However,some fruit- and nut-trees such as jackfruit,tamarind and coconut can be planted as living fences and as shelter for other crops.

Fig. Shelter for young trees

FACTORS IN TREE SELECTION

When selecting seedling or grafted varieties of fruit-trees for a home garden,study the characteristics of the tree's parents.Always choose healthy-looking trees with straight roots if buying from a nursery.Some things to consider are:

Time of harvest season. Is there fruit all year or only once a year'? Is this a time when other food is plentiful or in short supply'?

Size,taste,texture and use of fruit. Will the tree suit the conditions in your garden'? Can you grow crops under it or will the leaves block out too much light? *Tree shape and size.* Does the variety have strong branches or do they hang down and put the fruit too close to the ground? Will the fruit be easy to harvest? - *Disease and pest resistance.* Find out if there are any pests and diseases and how to manage them.Choose tree varieties which are known to be resistant to local pests and diseases.

PROPAGATION

Propagating fruit- and nut-trees requires special skills and experience and is best left to farmers who have fruit-tree nurseries.Other farmers can buy trees from the nursery after studying the characteristics of each variety.Buying trees reduces the risk and delay involved in growing your own.The best trees to buy are carefully selected and grafted,which means that they will be true copies of their mother plant.Grafted trees or trees grown from cuttings often have special names for each variety.For example,some well-known mango varieties are Kayu Manis (Indonesia),Nam Dorkmai (Thailand) and Carabao (Philippines).

PLANTING

Taking special care when planting seedlings or grafted trees will help them to establish quickly and safely.Tree roots should never be left exposed to sunlight or left where they will dry out.

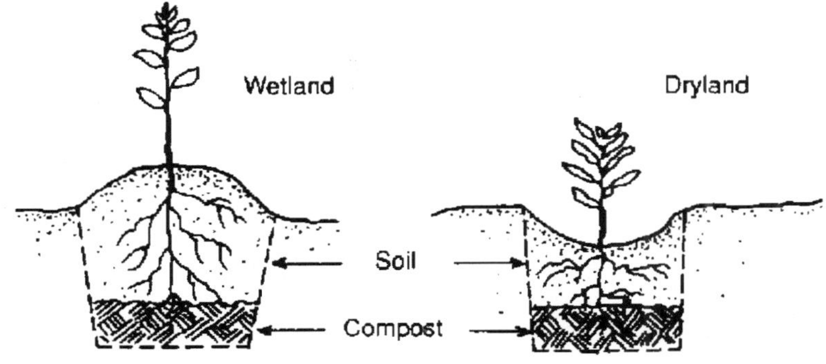

Fig. Tree planting

Dig a hole almost twice as deep as the length of the roots on the seedling and mix a generous amount of compost and a little fertilizer into the soil at the bottom of the hole.While holding the plant in the middle of the hole,fill the hole with soil and more compost.If the area is wet,plant the tree in a mound of soil higher than the general soil surface.If the area is dry,plant the tree in a basin,lower than the surrounding soil.

SPACING BETWEEN TREES

Plant trees with enough space between them to minimize competition.Look at a mature specimen of the tree you want to plant to get an idea of the spacing.For example,citrus tree's branches spread 1.5 m from one side to the other. This kind of citrus tree should,therefore,be planted with at least 1.5 m spacing.Many fruit-trees develop feeder roots close to the soil surface which can compete with other crops,so trees should be planted further apart when crops will be planted underneath.

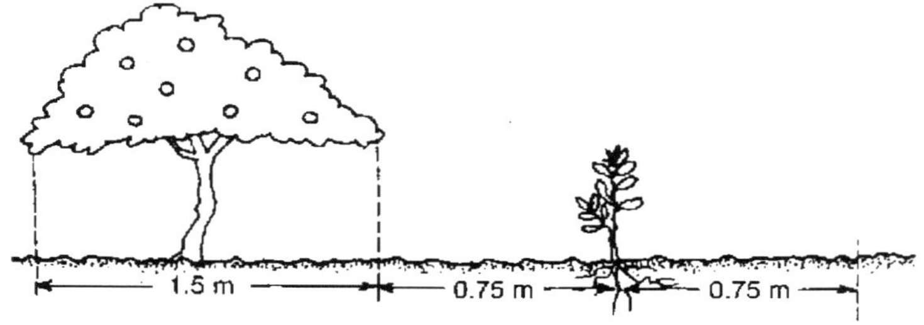

Fig. Tree spacing

PEST AND DISEASE CONTROL

Fruit-trees will be best prepared to fight pests and diseases if they are planted in conditions which suit them.Conditions include optimum light or shading,shelter,drainage and soil type.Many problems can be avoided if good farm hygiene is practiced: plant only healthy trees,remove and compost fallen or diseased fruit,prune out dead branches and do not bring soil from around infected plants into the garden.

TREK HUSBANDRY

Fruit-trees,like other plants and animals,will grow and produce better if they receive proper care.

Pruning. Some trees such as citrus,mango and sugar apple benefit from tree shaping. At planting,select the strongest upright branch to become the future trunk of the tree. As the tree grows, carefully prune out branches that are very close or rubbing together.

This lets air and light circulate through the tree,reduces diseases and can improve fruiting. Prune off weak branches and those that let fruit hang too close to the ground where animals or soil diseases can attack them. Take out dead branches where pests may be living.

- *Feeding.* Trees benefit from fertilizer,particularly at planting.Generally,a small handful of NPK fertilizer or 2 kg of good compost should be applied at planting.Every four months,apply some more.Apply fertilizer or compost before (not during) tree flowering and again when fruit is half-mature.Laying organic matter or mulch under the tree will help to provide organic matter,reduce weed competition and retain soil moisture.

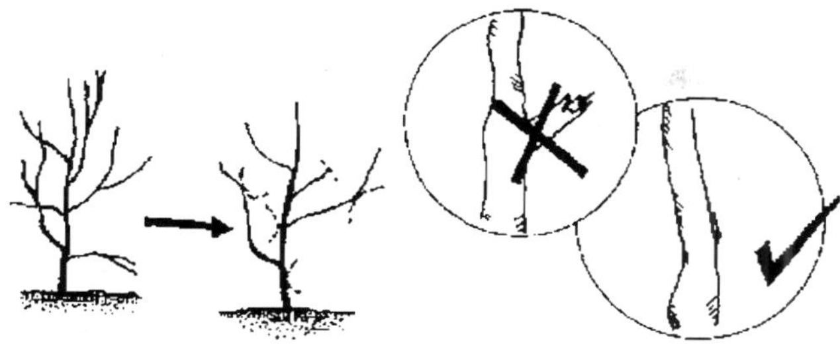

Fig. Pruning

- *Watering*.Young fruit-trees are sensitive to drought and need daily watering in the dry season for the first year or two.Older trees will be more resistant.Fruit such as papaya will benefit from daily watering for their whole life.

Table. Suggested Treks; Fruits, Nuts And Spices

Avocado	*Persea americana Mill.*
Banana	*Musa acuminata*
Breadfruit	*Artocarpus altilis (Parkins.) Fosb.*
Carambola, star fruit	*Averrhoa carambola L.*
Cashew nut	*Anacardium occidentals L.*
Citrus (many kinds)	*Citrus spp.*
Cocoa	*Theabroma cacao L.*
Coconut	*Cocos nucifera L.*
Coffee	*Coffea arabica, C. robusta*
Custard apple, sugar apple	*Annona reticulate, A. squamosa L.*
Durian	*Durio zibethinus L.*
Guava	*Psidium guajava L.*
Jackfruit	*Artocarpus heterophyllus Lam.*
Litchi	*Litchi chinensis Sonn.*
Mango	*Mangifera indica L.*
Papaya	*Carica papaya L.*
Passionfruit	*Passiflora edulis Deg.*
Pill nut	*Canarium ovatum Engl.*
Pineapple	*Ananas comosus Merr.*
Rambutan	*Nephelium lappaceum L.*
Rose apple	*Eugenia jambos L.*
Salak	*Salacca edulis Reinw.*
Sapodilla, chico	*Achras zopota L.*
Soursop	*Annona muricata L.*
Tamarind	*Tamarindus indica L.*

HOME GARDEN TECHNOLOGY: HOME GARDEN NURSERY

THE HOME GARDENS THE BEST PLACE FOR A NURSERY

A nursery in the home garden can be used to grow seedlings for all parts of the family farm areas.Locating the nursery within the home garden and close to the homestead means that seedlings and cuttings can receive regular watering and protection from pests and weed competition.For some plants,transplanted seedlings are more likely to survive in the field than those grown from directly sown seed.A nursery should always be well fenced to keep out animals.

Table. USES OF A NURSERY

Plant type	Activity
Vegetables	Grow seedlings in a seed bed before transplanting
	Grow roots on cuttings before transplanting
Fruit-trees	Grow seedlings
	Graft seedlings
Estate tree crop	Grow seedlings to a safe size for planting out

SEED BEDS AND SOWING

Large seeds such as beans and pumpkin are planted directly into the garden,but small seed such as lettuce,celery,tomato,coriander and rape should be germinated in seed beds and later transplanted to the garden bed.

The seed bed soil should be fine,with all sticks and stones taken out,and it should be raised about 15 cm above the surrounding area.Mix some sand into the soil to improve drainage and avoid seedling attack from fungus.Press the soil down hard with a flat board,then make shallow furrows in the soil.Sow the seed into the furrows then lightly cover the seed with soil.Make the soil firm by putting the board on the seeds and standing on it.Finally,put a thin mulch on the soil and water the seed bed.

Fig. A seed bed

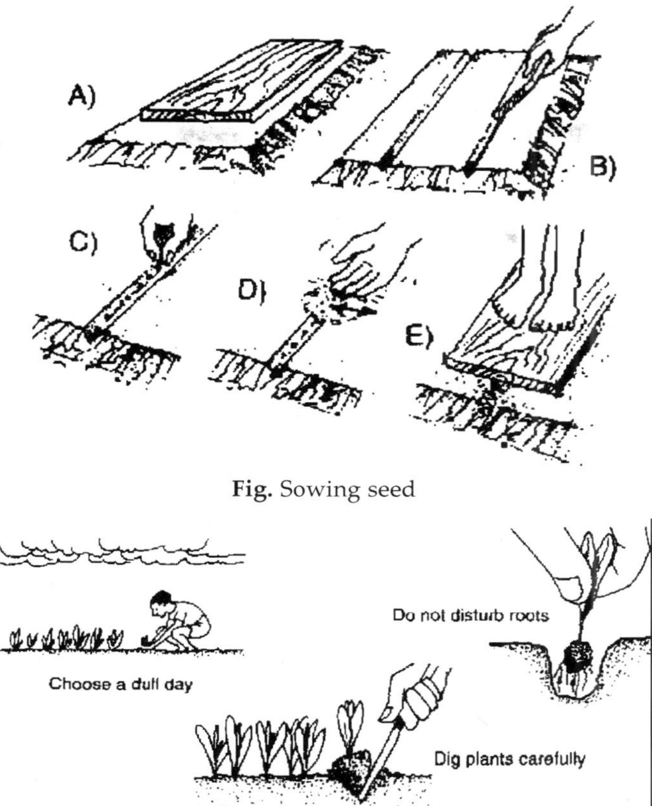

Fig. Sowing seed

Fig. Transplanting vegetable seedlings

SEED PRODUCTION

The quality of seed determines the success of a vegetable crop.Hybrid seeds need high inputs.Seed quality deteriorates over a few generations so,every five years,change the source of seed to one outside your home garden.Only choose the best plants to keep for seed and remove any diseased or poor plants before flowering so that their bad characteristics are not inherited in the seed.Dry seed well (but do not overdry it) and store in airtight bags or containers protected from rats.

CUTTINGS

Growing plants from parts other than seed,such as stem or root pieces,is one way to make new plants.These cuttings should be taken from the best plants and kept in the nursery until new roots or shoots form (two to four weeks).Use a sharp knife to make your cuttings.Cuttings always need sandy soil for good drainage but they must be watered regularly to avoid their drying out.Plant them in a seed bed.

Table. Some plants obtained from cuttings

Plant	Part of plant
Sweet potato, water spinach, taro	15-20 cm vine cutting top of root with shoot buds
Seedless breadfruit	25 cm root, 2 cm thick, set almost flat in the seed bed with top end 3 cm above the soil
Pineapple	Side shoots
Katuk (Sauropus sp.)	15 cm stem cutting, set 5-10 cm in the soil, with leaves above the ground

CONTAINER GROWING

Plants with expensive seeds and those that take a long time to grow (such as oil palm) can be grown in containers such as poly bags or pots made from strips of banana leaf.Make sure the pot is big enough for the plant's roots to grow without becoming cramped.The plant will transplant better if the roots have been allowed to grow long and deep.Soil in the containers should include some compost or a little fertilizer to feed the growing plant.

3

Soil and Fertilizer

MAINTENANCE OF SOIL FERTILITY

No two soils are alike either in respect of their nature or in respect of quantities of plant nutrients they contain.Under a given situation,the system of farming,soil management and manuring practices,etc.,influence the productiovity of soils and crop yields obtained from them. It is estimated that the different agricultural crops in India remove about 4.27 million tonnes of nitrogen,2.13 million tonnes of phosphoric acid,7.42 million tonnes of potash and 4.88 million tonnes of lime per year.The production of larger yields through improved varities of crops and intensive cultivation will increase the depletion of nutrients still further.But erosion and leaching cause additional losses.

The present production of synthetic nitrogenous fertilizers in the country reached 1.5 million tonnes of nitrogen,and the bulky organic manures might supply another 1.5 million tonnes of total nitrogen.The amounts of phosphorous,potash,etc.,added to the soil are very small.It is thus obvious that the current huge drain on nutrient supplies will continue to impoverish the soils unless these supplies are replenished by natural or by artificial means,the principal methods of supplementing natural recuperation and for improving the productive capacity of the soils are:

- To add organic matter to the soil,so that through decay,it may furnish a more or less continuous supply of nuttrients for crops,
  To restore or increase the amount of defficient nutrients by the application of fertilizers.

The urgent need of the constantly expanding agriculture production to meet the requirements of the continually increasing human and cattle populations in India makes the supply of additional plant nutrients through fertilizers and organic manures,a problem of supreme importance.

TYPES OF MANURES AND FERTILIZERS

Indian soils are usually very poor in organin matter as well as in nitrogen.Pohsphate deficiency is less wide-spread and potash deficiency generally occurs in com-pact areas.In acid soils the addition of lome steps up production.Materials which are commonly used to maintain and improve soil fertility

- *Manures*: These are relatively bulky materials,such as animal or green manures,which are added mainly to improve the physical condition of the soil,to replenish and keep up its humus status,to maintain the optimum conditions for the activities of soil micro-organisms and make good a small part of the plant nutrients removed by crops or otherwise lost through leaching and soil erosion.
- They,thus,supply practically all the elements of fertility which crops require,though not in adequate proportions.The plant-food elements contained in a manure are released in an available form after it is applied to the soil and is decomposed by soil micro-organisms.Similarly,the green manures add not only substantial amounts of organic matter but also nitrogen.
- *Fertilizers*: Fertilizers are inorganic materials of a concentrated nature; they are applied mainly to increase the supply of one or more of the essential nutrients,e.g.nitrogen,phosphorous and potash.Fertilizers contain these elements in the form of soluble of readily available chemical compounds.This distinction is,however,not very rigid.In common parlance the fertilizers are sometimes called 'chemical','artificial' or 'inorganic' manures.
- *Concentrated Organic Manures*: Some of the concentrated materials,such as oil-cakes,bone-meal,urine and blood are of organic origin.The use of manures and fertilizers is complementary and not as a substitute for each other.
- *Bulky Organic Manures*: The properties and role of organic matter and humus in the soil have been explained already.

Farmyard Manure

Good-quality farmyard manure is perhaps the most valuable organic matter applied to a soil.It is the most commonly used organic manure in India.It consists of a mixture of cattle dung,the bedding,used in the stable and of any ramnants of straw and plant stalks fed to cattle.

Though its crop-increasing value has been recognised from time immemorial,more than 50 per cent of the cattle dung produced in the country today is burnt as fuel and is thus lost to agriculture.Not only this tremendous waste,but also the tradition method of preparing and storing the farmyard manure is generally faulty.

The cattle-dung,together with stable-waste and house sweeping,is first collected in the open backyard,and when a cartload has been collected,it is removed to another heap or to an uncovered pit in a common plot outside the village.The loose heaps lie exposed to the sun,with the result that the raw organic matter dries up quickly and does not rot properly. Very often,a part of the dry dung is blown off by wind or washed away by rain.Cattle urine is either not conserved or is stored in a defective manner.American studies on the distribution of soil derived elements between urine and faeces of dry cows have shown that 95 per cent of Potassium,63 per cent of nitrogen and 50 per cent of sulphur are contained in the urine.

The wastage of nitrogen-rich urine,the loss of nitrogen(in the form of ammonia) due to the fermentation of exposed cattle dung,and the washing away of soluble mineral elements be leaching reduce its manurial value in India to a great extent.Its average content of plant nutrients under Indian and European conditions is shown below for comparison:

Percentage Content

	N	P2O5	K2O
India	0.3	0.15	0.3
European countries	1.0	0.30	1.0

About half of this nitrogen,one-sixth of phosphorous and more than half of potash are readily soluble and subject to dissipation.However,the loss of nitrogen and minerals elements caused by careless handeling can be reduced greatly by using absorbent bedding for cattle,storing dung in stone or brick-line pits,mixing large quantioties of straw and other vegitable matter with cattle dung,and keeping the heap compact and moist.

Thus,if urine is properly conserved,the loss of soluble mineral elements through seepage is prevented,bacterial decomposition of raw organic matter is encouraged,plant nutrients are made soluble,and nitrogen losses are minimised.

Table.The Relative Absorbent Capacity of Different Materials used as Bedding for Cattle

Material following	**Quantity of water (in kg) retained by one kg of the materials after 24 hours of soaking**
Wheat straw	2.20
Peat straw	2.80
Dry leaves	2.00
Peat	6.00
Sawdust	4.35
Soil	0.50
Sand	0.25

If urine is not conserved in the bedding used for cattle it must be collected in covered *pucca* cistern and then,added to the dung in the manure pit.Nitrogen in the urine is mainly in the form of urea,which readily changes into the highly volatile ammonium carbonate through bacterial action,and quickly loses ammonia thereafter by evapouration.This loss can be reduced to a great deal if the manure and the urine-soaked absorptive itter for bedding are kept compacted in a pit. The pit may be 1 m in depth,1.3 to 1.5 m in width and 4.5 to 6 m in length,depending upon the no.of cattle on a farm.The filling of the pit should be 'sectional' and when each section of three or 1.3 m in length is filled to about 45 cm above the ground level,it should be clustered with 2.5 cm layer of a mixture of mud and dung in equal proportions.

Before plastering,4 to 5 buckets of water should be added to the manure in the pit.Plastering conserves moisture and nitrogen and also prevents housefly nuisance.The manure becomes ready for use in about 4 to 5 months after plastering. The quality of manure is also improved by the concentrated feeds given to cattle.Cotton-seed,cotton-seed cake,linseed-meal,wheat bran,grain husk,groundnut cake,gram,horse-gram,etc.are rich in nitrogen, phosphorous, potassium, magnesium and sulphur.

It has been found that in the case of adult working-cattle about 80 per cent of nitrogen and the other mineral elements contained in the feed is recovered in urine,faeces and other animal by-products.Accordingly,manure from cattle fed on cereal straws and grass hay is much less valuable than that from animals fed on legume hays,grains and concentrates.

In foreign countries,considerable attention has been given to the use of preservatives on manure.Calcium sulphate or gypsum and superphosphate have proved most promising in preventing the escape of ammomnia.Gypsum has been found specially effective as an ammonia-absorbing agent. Superphosphate, besides absorbing ammonia,supplies additional phosphorous and,thus,improves the crop producing capacity of the manure.

Partially rotten farmyard manure should generally be applied to the soil about three to four weeks before the sowing of a crop.In case there is sufficient moisture in the soil,there sill be enough time for its decomposition and for improving the soil structure.Its application too long before sowing a crop will either cause a drying up of the rotted manure or too quick decomppostion, depending on the incidence of rains.But in each case,there will be a serious loss of ammonia and nitrogen.If the manure is already well rotted,it is advisable to apply it just before sowing to a crop.

This procedure is perticularly essential in the case of light soils.In any case,after the manure is carted to the field,it should be evenly spread and worked into the soil soon to avoid the loss of nitrogen.The existing practice of leaving the manure in small heaps scattered in the field for several days,before spreading it in the field and incorporating it into the soil,results in the serious deterioration of its quality,particularly if strong winds blow.

In vagitable and fruit cultivation,the application of well-rotten manure,in conjunction with fertilizers to young plants individually,has been founnd to give the best results.In Egypt,even the cotton crop,which is invariably grown on ridges,is given a top-dressing a handful of the rotten manure being applied to the soil at the base of each plant before an irrigation,and is worked into the soil with a hand hoe. The practice of penning cattle,sheep and goats in the fields in summer is common in some parts of the country.Folding 7,000 sheep for one night is said to add the equivalent of 149.3 quintals(14.93 tonnes) of cattle dung.The fresh dung left in the field in such cases rapidly dries up.This drying checks ammonification and loss of nitrogen.With the first fall of rain,the dung is worked into the soil.It,therefore,does not lose much of its fertilizing value.Further more,its beneficia effects on the physical condition of the soil are undeniable.However,sheep-folding is said to make the land more weedy.

There must be adequate moisture in the soil for the proper decomposition of organic matter.Farmyard manure can,therefore,be applied to all crops grown in the rainy season or grown under irrigation.The quantity of manure to be applied to unirrigated crops varies from 1.5 to 2 cartloads per hectare in areas of heavier rainfall.

If sufficient farmy manure is not available,it may be applied at the usual rate to a part of the land,say,to one-third or one-forth of the area,in rotation every year,so that all parts of the field receive the manure regularly once in three or four years.For irrigated field crops,the rate varies from 10 to 20 cartloads.Sugarcane,maize and garden crops,such as potatoes, turmeric, ginger, vegetables and fruits receive still higher doses,amounting sometimes to 15 to 25 cartloads.A cartload of manure,measuring 9 cubic metres,weighs about half a tonne.

It must be stressed that the value of farmyard manure in soil improvement is due to its content of principal nutritive elements and its ability to:

- Improve the soil tilth and aeration,
- Increase the water-holding capacity of the soil,
  Stimulate the activity of micro-organisms that make the plant-food elements in the soil readily available to crops.

The supply of organic matter,which is later converted into humus is a property of farmyard manure.

One tonne of carttle dung can supply only 2.95 kg of nitrogen,1,59 kg of phsophoric acid and 2,95 kg of potash.

The use of farmyard manure alone causes an imbalance in nutirtion owing to its relatively low content of phosphoric acid.Therefore,to keep the soils well supplied with all the essential elements of plant food in a readily available form,and also to keep them in good 'heart',it is advisable to use the bulky organic manures in conjuction with superphosphate and such other artificial fertilizers as contain the particular plant food or foods in which a soil may be deficient or which the crop to be grown may specially require.

SOIL FERTILITY AND SOIL MINERALOGY

Soil mineralogy is closely related to soil fertility.Differences in soil mineralogy cause great differences in soil fertility.For instance,moderately weathered clays attract and retain greater amounts of nutrients than highly weathered clays and oxides.Though volcanic ash inherently has a low ability to hold nutrients,it interacts with organic matter to produce very fertile soils.Knowledge of mineralogy helps to determine the appropriate nutrient management strategy for your soil.

In our discussion on soil texture and structure,we mentioned that the very small particles form aggregates.The major groups of small particles include silicate clays,organic matter,volcanic ash minerals,and oxides.In fact,this grouping of small particles is also used to describe the distinct categories of soil mineralogy (types of clay minerals).

Soil mineralogy:

- Layered silicate clays
  - High activity
  - Low activity
- Organic matter
- Volcanic ash material
- Oxides

Soil behaviour is greatly influenced by these types of clay minerals and/or the amount of organic matter that a particular soil type contains.

Before discussing differences in soil mineralogy,it is very helpful to understand the concepts of cation and anion exchange capacities (CEC and AEC).CEC and AEC are properties that can help differentiate soil minerals.

Cation Exchange Capacity (CEC) and Anion Exchange Capacity (AEC)

CEC and Less Weathered Soils

Less weathered soils,that contain minerals such as montmorillonite,are said to have a 'cation exchange capacity,' or CEC,under acidic,neutral,and alkaline conditions.CEC is the soil's ability to attract,retain,and supply nutrients,such as calcium,potassium,ammonium,and magnesium.These nutrients are positively charged atoms,known as cations.The surfaces of less-weathered clay minerals (such as montmorrillonite) generally have a negative charged.Much like a magnet,negatively charged soil surfaces attract positively charged cations.However,under acidic conditions,the soil will also have a tendency to attract aluminum and hydrogen cations.The presence of aluminum and hydrogen contributes to soil acidity.In contrast,under alkaline conditions,the soil attracts sodium which contributes to soil alkalinity.

AEC and Highly Weathered Soils

Although the majority of the world's soils have CEC,the highly weathered soils of the tropics are an exception.In addition to the having CEC,many tropical soils also have an 'anion exchange capacity,' or AEC,depending upon the pH of the soil.Under neutral and alkaline conditions,the soil has CEC,like the less weathered soils.However,under acidic conditions,these soils generate AEC.This means that the soil becomes positively charged and attracts,retains,and supplies negatively charged anions,such as sulfate,phosphate,nitrate,and chloride.For soils with AEC,proper management of pH is crucial in order to provide sufficient amounts of the nutrient cations (calcium,magnesium,ammonium,and potassium).

Minerals that exhibit AEC are highly weathered kaolinite,aluminum and iron oxides,organic matter,and the allophanes and imogolites of volcanic soils.Highly weathered Ultisols and Oxisols,volcanic Andisols,and organic Histosols all have AEC under acidic conditions.The pH at which these soils develop AEC differs depending upon the minerals within the soil.Since organic matter only generates AEC at a very low pH,it is still a good source of CEC.

Importance of CEC and AEC

CEC and AEC values are important measurements that provide us with important information regarding the soil's ability to retain and supply certain nutrients to the plant.In addition to nutrient retention,CEC and AEC helps us predict the leaching potential of certain nutrients in areas with high rainfall.When the soil has a very high CEC,negatively charged nutrients such as nitrate are not be retained by the soil.Instead,nitrate leaches through the soil profile in areas with high amounts of precipitation.Likewise,soils with high AEC experience leaching of positively charged nutrients,such as calcium and potassium.

CEC is often expressed in centimoles of charge per kg of soil.By sending a sample of your soil to a soil testing laboratory,you can determine your soil's CEC.The value of knowing a soil's CEC cannot be underestimated in nutrient management.Without it,your soil would not be able to provide your plants with sufficient amounts of nutrients.However,one must be cautious when interpreting the laboratory results for CEC depending on the soil type being analysed.You can obtain accurate results for most laboratory methods that measure CEC in less weathered soils with permanently negative charge.On the other hand,these laboratory methods can overestimate the CEC of highly weathered soils that have an AEC.This is because the pH of highly weathered soils affects the CEC of the soil.As the pH of highly weathered soils with AEC increases,the CEC of the soil also increases.Thus when the method used to determine the CEC uses solutions that raise the pH of the soil,the reported CEC is higher than its actual value in the field.

Layered Silicate Clays

Layered silicate clays are secondary minerals that have formed as the result of weathering of parent material.There are two major categories of layered silicate clays within the soil: high activity clays and low activity clays.

High Activity Clays

Generally,soils with large amounts of high activity clays are not highly weathered.High activity clays have a high 'cation exchange capacity' (CEC),due to their large surface area.This means that these clays have a great capacity to retain and supply large quantities of nutrients,such as calcium, magnesium, potassium, and ammonium.Not only do these clays have a large CEC,but they will generate CEC under all soil conditions regardless of soil pH.As a result,these clays tend to produce highly fertile soils.Examples of these clays are montmorillonite (and other smectites),vermiculite,illite,and mica.

Cation Exchange Capacity or Anion Exchange Capacity

- High activity clays have a cation exchange capacity (CEC).Although high activity clays will not have anion exchange capacity (AEC),the CEC increases as pH increases and decreases as pH decreases.

Montmorillonite

Some high activity clays,such as montmorrillonite,have a shrink and swell potential.This means that the clays will shrink and crack when dry,and expand and swell when wet.With little additions of nutrients,these soils may be very productive.

However,the shrink and swell potential will result in poorer drainage.And so,proper management of irrigation is required.

Low activity clays

In contrast,low activity clays are more highly weathered.Thus,due their lesser surface area,low activity clays have a lower capacity to retain and supply nutrients.In addition to CEC,low activities clays can also have AEC,depending upon the pH of the soil.

The AEC causes these clays to retain and supply nutrients,such as phosphate,sulfate,and nitrate,rather than the base cations,under acidic conditions.Yet,under neutral and alkaline conditions,these low activity clays generate a CEC.

Cation Exchange Capacity or Anion Exchange Capacity?

- Under acidic conditions,low activity clays have an AEC
- Under neutral and alkaline conditions,low activity clays have a CEC

Management of pH

In order to provide adequate amounts of the base cations,proper management of pH is crucial.If soil pH is low,additions of lime and/or organic matter may increase the CEC for soils high in low activity clays.

Low activity clays have a low shrink and swell potential.With additions of nutrients,these soils may be very productive soils.

Table. CEC and Surface Area of Common Soil Minerals

Mineral	Type	CEC (surface charge cmolc/kg^{-1})	Surface area (external m^2/g^{-1})
Smectite	High activity clay	-80 to -150	80 to 150
Vermiculite	High Activity clay	−100 to -200	70 to 120
Fine Mica	High activity clay	−10 to -40	70 to 175
Chlorite	High activity clay	−10 to -40	70 to 100
Kaolinite	Low activity clay	−-1 to -15	5 to 30
Gibbsite	Al-oxide	+10* to -5	80 to 200
Goethite	Fe-oxide	+20 to -5	100 to 300
Allophane	Amorphous	+10 to -150	100 to 1000
Humus	Organic	−100 to -500	Variable

Organic Matter

Most soil organic matter accumulates within the surface layer of the soil.This organic matter may be divided into two groups: non-humic matter and humic matter.

Non-humic matter includes all undecomposed organic material within the soil.Examples of non-humic matter are twigs,roots,and living organisms.Humic matter includes humic acids,fulvic acids,and humin.(Humin is the dark material in soil that is highly resistant to decomposition.)

Importance of soil organic matter

- Due to its tremendous surface area,soil organic matter:
  - Acts like a sponge to store water
  - Retains and provides nutrients (CEC)
  - Glues and binds soil particles into stable aggregates
- Reduces the occurrence of aluminum toxicities.

Like low activity clays,organic matter may have either CEC or AEC,depending upon soil pH.However,it will rarely have AEC.In fact,the pH must fall to approximately 2.0 before it will have AEC.

Cation Exchange Capacity or Anion Exchange Capacity?:

- Soil organic matter may have both AEC and CEC.However,the charges on organic matter are dependent upon soil pH.For soil organic matter to generate an AEC,the soil pH must be 2.0.

Management

Without additions of organic matter,tillage practices will greatly reduce organic matter content in the soil.And so,no-till and minimum tillage systems with the return of organic matter to the soil are gaining favour by farmers to improve and conserve soil quality.

BIODIVERSITY AND FARMING OF SOIL

Biodiversity is an important aspect of the relatively new discipline of ecology,which plays a major role in sustainable development.Involving more than the problem of loss of species that appears to be the common concern,biodiversity refers to all species of plants,animals and micro-organisms that exist and interact within an ecosystem.When species are lost,so are certain functions and developments up and down the hierarchical scale.In natural ecosystems,the vegetation cover of a forest or grassland prevents soil erosion,replenishes groundwater and controls flooding by enhancing infiltration and reducing water run-off.There is interaction between the soil and the rest of the ecosystem. In agricultural systems,biodiversity performs ecosystem services beyond production of food,fibre,fuel,and income.Examples include recycling of nutrients,control of local microclimate,regulation of local hydrological processes,regulation of the abundance of undesirable organisms,and detoxification of noxious chemicals.

These are mainly ecological processes and to maintain them a certain level of functional biodiversity is required.From an economic point of view,this means that if an agricultural system is deprived of basic regulating functional components,resulting in loss of soil fertility and pest and disease regulation,there is an increasing need for costly external inputs.Swift and Anderson stated that the net result of biodiversity simplification for agricultural purposes is an artificial ecosystem that requires constant human intervention,whereas in natural ecosystems the internal regulation of function is a product of plant biodiversity through flows of energy and nutrients.This form of control is progressively lost under agricultural intensification.

This means that enhancing biodiversity in agro-ecosystems is a key strategy to bring sustainability to production.This involves a shift towards mature or climax characteristics,a move away from an over-emphasis on production (developmental phase) towards more quality and internal regulation.

Agro-ecological-based farming systems like organic farming should include strategies that exploit the complementarities and synergies resulting from various combinations of crops,trees and animals.These include arrangements such as poly-cultures,agro-forestry systems and crop-livestock mixtures.Insect pest problems are increasingly linked to the expansion of mono-cropping and the loss of landscape diversity or local habitat diversity.Odum concluded that with the maturing of a system the biodiversity increases,which is accompanied by a reduction in insect pests,increased stability,less losses from the system,and so on.

SOIL-PLANT-ANIMAL HEALTH AND ORGANIC SOIL MANAGEMENT

The view that plant health,animal health and human health are related to soil conditions has been promoted by many scientists with a strong academic background,such as Steiner,Howard,Balfour,Muller and Rusch, Boucher, Pfeiffer, Albrecht,Voisin,Hamaker and later Chaboussou.The link between plant health and soil conditions is becoming increasingly obvious in modern research.Chaboussou's work leads him to the following statement: "Healthy plants do not get sick".

I. Disease suppression in plants as a function of the soil (life) has been discussed at various conferences.
II. Chabossou indicates that metabolic disturbance in plants caused by pesticides influences or unbalanced crop nutrition influences plant health negatively.

Trace-elements and other mineral balances in the soil are also becoming increasingly recognised as having an influence on plants and animal health.Condron et al.stated "Trace element deficiencies occur in the human population in New Zealand as well as in livestock".Furthermore,Condron et al.,recognised the potential of organic farming systems to produce plant and animal products appropriate for the human diet.Voisin stated that the soil must be kept in good health if the animal is to remain in good health.The same is true of man.Soil Science is the foundation of protective medicine,the medicine of tomorrow.

A UNDP on sustainable farming states: "In reality organic agriculture is a consistent systems approach based on the perception that tomorrow's ecology is more important than today's economy".

This report advocates readjustment by economy to primary production factors and not the other way around.Without ecology,there is no economy.If conventional agriculture had been made to pay for the degradation and environmental damage it caused,the move towards ecological and organic farming systems would have been made long ago.The aim is to stop degradation and re-establish natural balances.In Europe,research emphasis has shifted from the validation of the organic farming system to finding practical solutions for theoretical and especially practical challenges.In organic farming there is a different,organised 'practice' with its own ethos,which ought to have consequences for scientific research.An ecological or organic scientific platform could close the gap between 'grass roots' organic organisations and policymakers and other research institutions.The dependence of organic farms on biological activity (for instance,to supply nutrients to plants) requires a different methodical systems approach in which a mutual transfer of knowledge between farmer and scientist is valued.

Yield and production are seen in the local literature as limiting factors for the development of New Zealand organic farming.There is a different emphasis in organic farming systems,an emphasis on quality rather than quantity,on optimum production rather then on maximum production.However,both can be significantly improved (quantity and quality) if the same amount of research funding was allocated to researching organic farming systems as to mainstream farming systems.IPM strategies (a strategy between conventional and organic farming), for example,receive more funding than organic farming systems.These strategies,however,involve the first two steps towards sustainability: improving efficiency of existing systems and input substitutions.The third step is generally not taken outside the organic sector: a conscious re-design of the agro-ecosystem.This step requires knowledge and understanding of those agro-ecosystems.

In Europe,dedicated organic or ecological research centres have gathered this knowledge and understanding and are defining and researching the conscious re-design of agro-ecosystems in conjunction with farmers.In New Zealand such understanding is mainly held by a few experienced organic farmers and scientists.For scientists to help more farmers adopt truly sustainable organic farming systems,wider understanding of and research into soil ecological processes is required.

FERTILIZERS AND NUTRIENT MINING IN SOILS

The only answer to the "nutrient mining" problem is to bring fresh nutrients into the area from some richer source,and fortunately natural processes have concentrated the major nutrients in locations around the world.All "synthetic" nitrogen fertilizers obtain their nitrogen from the atmosphere.This is "fixed" using the *Haber process*,in which dihydrogen and dinitrogen at high temperature and pressure are combined over a catalyst,which speeds up the reaction forming ammonia:

$$N_2(g) + 3H_2(g) \rightarrow 2NH_3(g).$$

The hydrogen required for the Haber process is normally made using methane separated from natural gas.Methane is reacted with water and air at high temperatures over catalysts to form a mixture of dihydrogen,dinitrogen (from the air) and carbon dioxide.Carbon dioxide and water are removed (the carbon dioxide is used later to make urea),leaving dihydrogen and dinitrogen in the correct proportion to combine to form ammonia (unbalanced equation)

$$CH_4(g) + H_2O(g) + [O_2(g) + N_2(g)] \rightarrow [N_2(g) + 3H_2(g)] + CO_2(g).$$

The ammonia obtained from the Haber process is converted into a form more useful as fertilizer.It can be combined with carbon dioxide to form urea,which is chemically identical to the urea excreted in the urine of animal:

$$2NH_3(g) + CO_2(g) \rightarrow O{=}C(NH_2)_2(s) + H_2O(l)$$

or combined with sulfuric,nitric or phosphoric acid to form ammonium sulfate,ammonium nitrate or diammonium hydrogen phosphate $(NH_4)_2(HPO_4)$.The nitric acid is formed by oxidising ammonia over a catalyst

$$NH_3(g) + 2O_2(g) \rightarrow HNO_3(g) + H_2O(l)$$

Phosphorus

Phosphorus is a relatively abundant element in the earth's crust,and occurs naturally in marine deposits of *phosphate rock*,as calcium phosphate,$Ca_3(PO_4)_2$,or the minerals *apatite*,$CaF_2.3Ca_3(PO_4)_2$,or *hydroxy apatite*,$Ca(OH)_2.3Ca_3(PO_4)_2$ (hydroxy apatite is the mineral that forms bones and teeth).The phosphate rock from the Pacific region is found on coral islands where flocks of sea birds roost.It is thought that the phosphate in their excreta,derived from their fishy diet,combines over time with coral to form the phosphate rock.Some phosphate rock is sufficiently soluble so that the phosphorus from the finely ground rock is available to plants.Other phosphate rock is too insoluble for direct use,and is commonly treated with sulfuric acid to make a mixture of calcium sulfate and the more soluble calcium hydrogen phosphate,called *superphosphate*.

$$Ca(OH)_2.3Ca_3(PO_4)_2 + H_2SO_4 \rightarrow CaSO_4 + Ca(HPO_4)_2$$

Phosphate rock can also be reacted with excess sulfuric acid to make phosphoric acid,H_3PO_4,which can be converted into ammonium hydrogen phosphate,$(NH_4)_2(HPO_4)$ for use in soluble fertilizers,or into sparingly soluble magnesium ammonium phosphate $Mg(NH_4)(PO_4)$ (magamp) for use as a slow-release fertilizer.

Sulfur

Sulfur is also an essential nutrient,and some areas of New Zealand are sulfate deficient.Use of superphosphate,or ammonium,potassium or calcium sulfate (gypsum) as fertilizer supplies the needed sulfate.The sulfuric acid needed to make superphosphate is obtained by burning sulfur to form sulfur dioxide,

$$S(s) + O_2(g) \rightarrow SO_2(g)$$

which is oxidised over a catalyst to form sulfur trioxide,

$$2SO_2(g) + O_2(g) \rightarrow SO_3(g)$$

which reacts with water to form sulfuric acid,

$$SO_3(g) + H_2O(l) \rightarrow H_2SO_4(aq).$$

The sulfur is obtained from deposits of elementary sulfur found in volcanic areas of the world,or is extracted from natural gas,which in some places contains hydrogen sulphide.Sulfur dioxide is released into the atmosphere from volcanoes.This is ultimately converted into sulfuric acid in the atmosphere; this is washed out by rain,so providing a "natural" source of sulphur.The 1995 eruption of Ruapehu "topdressed" much of the central North Island with economically useful amounts of sulphate.Sulfur dioxide is also released into the atmosphere when coal or oil containing sulfur is burned in power stations,and the resulting sulfuric acid in the atmosphere causes "acid rain" in some industrialised areas.

Potassium

Potassium is the other main nutrient needed,and this is obtained from sea water,directly or indirectly.When sea water evaporates sodium chloride (common salt) crystallises first.If the liquid remaining (bittens) is evaporated potassium chloride crystallises.In some places this occurred over geological times when oceans were isolated by geological processes and then evaporated,leaving large deposits of salt (which crystallised first) and of potassium chloride (which crystallised later).The potassium chloride can be used directly as a fertilizer,and is satisfactory for use at low rates,but this has the disadvantage that chloride is also added,and many plants are not very tolerant of high levels of chloride.

It is therefore usual to convert the potassium chloride to potassium sulfate (which also occurs in natural deposits) for use as a fertilizer.The so-called "un-natural" inorganic fertilizers are thus derived from the very natural sources of the air (plus natural gas as a hydrogen source) for nitrogen,coral rock plus sea-bird droppings (plus sulfur from volcanoes or "sour" natural gas to make superphosphate) for phosphorus,and the ocean for potassium.

THE GENERAL PRINCIPLES OF MANURING

In their natural condition all soils not absolutely barren are capable of supporting a certain amount of vegetation,and they continue to do so for an unlimited period,because the whole of the substances extracted from them are again restored,either directly by the decay of the plants,or indirectly by the droppings of the wild animals which have browzed upon them.Under these circumstances,a soil yields what may be called its normal produce,which varies within comparatively narrow limits,according to the nature of the season,temperature,and other climatic conditions.

But the case is completely altered if the crop,in place of being allowed to decay on the soil,is removed from it,for,though the air will continue to afford an undiminished supply of those elements of the food of plants which may be derived from it,the fixed substances,which can only be obtained from the soil,decrease in quantity,and are at length entirely exhausted.

In this way a gradual diminution of the fertility of the soil takes place,until,after the lapse of a period,longer or shorter,according to its natural resources,it will become entirely incapable of maintaining a crop,and fall into absolute infertility unless the substances removed from it are restored from some other source in the form of manure.

When this is done,the fertility of the soil may not only be sustained but greatly increased,and,in point of fact,all cultivated soils,by the use of manure,are made to yield a much larger crop than they can do in their natural condition.

The fundamental principle upon which a manure is employed is that of adding to the soil an abundant supply of the elements removed from it by plants in the condition best fitted for absorption by their roots; but looked at in its broadest point of view,it acts not merely in this way,but also by promoting the decomposition of the already partially disintegrated rocks of which the soil is composed,setting free those substances it already contains,and facilitating their absorption by the plants.

In considering the practical applications of the broad general principle just stated,it might be assumed that a manure ought invariably to contain all the elements of plants in the quantities in which they are removed by the crops,and that when this has been accurately ascertained by analysis,it would only be necessary to use the various substances in the pro this,though a very important,and no doubt in many cases essential condition,is by no means the only matter which requires to be taken into consideration in the economical application of manures.And this becomes sufficiently obvious when the circumstances attending the exhaustion of the soil are minutely examined.

When a soil is cropped during a succession of years with the same plant,and at length becomes incapable of longer maintaining it,the exhaustion is rarely,if ever,due to the simultaneous consumption of all its different constituents,but generally depends upon that of one individual substance,which,from its having originally existed in the soil in comparatively small quantity,is removed in a shorter time than the others.To restore the fertility of a soil in this condition,it is by no means necessary to supply all the different substances required by the plant,for it will suffice to add that which has been entirely removed.

On the other hand,if an ordinary soil be supplied with a manure containing a very small quantity of one of the elements of plant food,along with abundance of all the others,the amount of increase which it yields must obviously be measured,not by those which are abundant,but by that which is deficient; for the crop which grows luxuriantly so long as it obtains a supply of all its constituents,is arrested as effectually by the want of one as of all,as has been proved by the experiments of Prince Salm Horstmar and others,referred to in a previous chapter; and hence,in order to obtain a good crop,it would be necessary to use the manure in such abundance as to supply a sufficiency of the deficient element for that purpose.

If this course were persevered in for a succession of years,the other substances which would have been used in much more than the quantity required by the crops,must either have been entirely lost or have accumulated in the soil.In the latter case it is sufficiently obvious that the soil must have been gradually acquiring an amount of resources which must remain dormant until the system of manuring is changed.

To render them available,it is only necessary to add to it a quantity of the particular substance in which the manure hitherto employed has been deficient,so as to restore the lost balance,and enable the plant to make use of those which have been stored up within it.

The substance so used is called a *special* manure; that containing all the constituents of the crop is a *general* manure.The distinction of these two classes of manures is very important in a practical point of view,because a special manure is not by itself capable of maintaining the life of plants,but is only a means of bringing into use the natural and acquired resources of the soil.

In place of preventing or retarding its exhaustion,it rather accelerates it by causing the increased crops to consume more portions thus indicated.But this,though a very important,and no doubt in many cases essential condition,is by no means the only matter which requires to be taken into consideration in the economical application of manures.And this becomes sufficiently obvious when the circumstances attending the exhaustion of the soil are minutely examined.

When a soil is cropped during a succession of years with the same plant,and at length becomes incapable of longer maintaining it,the exhaustion is rarely,if ever,due to the simultaneous consumption of all its different constituents,but generally depends upon that of one individual substance,which,from its having originally existed in the soil in comparatively small quantity,is removed in a shorter time than the others.To restore the fertility of a soil in this condition,it is by no means necessary to supply all the different substances required by the plant,for it will suffice to add that which has been entirely removed.

On the other hand,if an ordinary soil be supplied with a manure containing a very small quantity of one of the elements of plant food,along with abundance of all the others,the amount of increase which it yields must obviously be measured,not by those which are abundant,but by that which is deficient; for the crop which grows luxuriantly so long as it obtains a supply of all its constituents,is arrested as effectually by the want of one as of all,as has been proved by the experiments of Prince Salm Horstmar and others,referred to in a previous chapter; and hence,in order to obtain a good crop,it would be necessary to use the manure in such abundance as to supply a sufficiency of the deficient element for that purpose.

If this course were persevered in for a succession of years,the other substances which would have been used in much more than the quantity required by the crops,must either have been entirely lost or have accumulated in the soil.

In the latter case it is sufficiently obvious that the soil must have been gradually acquiring an amount of resources which must remain dormant until the system of manuring is changed.To render them available,it is only necessary to add to it a quantity of the particular substance in which the manure hitherto employed has been deficient,so as to restore the lost balance,and enable the plant to make use of those which have been stored up within it.The substance so used is called a *special* manure; that containing all the constituents of the crop is a *general* manure.

The distinction of these two classes of manures is very important in a practical point of view,because a special manure is not by itself capable of maintaining the life of plants,but is only a means of bringing into use the natural and acquired resources of the soil.

In place of preventing or retarding its exhaustion,it rather accelerates it by causing the increased crops to consume more abundantly,and within a shorter period of time,those substances which it contains.On the other hand,a general manure prevents or diminishes the consumption of the elements of plant-food contained in the soil,and if added in sufficient abundance,may cause them to accumulate in it,and even enable an almost absolutely barren soil to yield a tolerable crop.

General manures must therefore always be the most important and essential,and no others would be used if it were possible to obtain them of a composition exactly suited to the requirements of the crop to be raised.

Practically,however,this condition cannot be fulfilled,because all the substances available for the purpose,and particularly farm-yard manure,are refuse matters,the exact composition of which is not under our control,and they do not necessarily contain their constituents either in the most suitable proportions,or the most available forms,and consequently when they are used during a succession of years,certain of their constituents may accumulate in the soil,and it is under such circumstances that special manures are both necessary and advantageous.

Several different substances,but more especially farm-yard manure,fulfil in a very remarkable manner the conditions of a general manure,and supply abundantly,not merely the mineral,but also the carbonaceous and nitrogenous matters necessary for building up the organic part of the plant; and hence its use is governed by principles of comparative simplicity,and really resolves itself into determining the best mode of managing it so as effectually to preserve its useful constituents,and,at the same time,to bring them into those forms of combination in which they are most available to the plant.

But the employment of a special manure opens up nice questions as to the relative importance of the different elements of plants which have given rise to much controversy and difference of opinion.In treating of the food of plants,it has been already observed that the fixed or mineral constituents which are contained in their ash,are necessarily derived exclusively from the soil,but that the carbon,hydrogen,nitrogen,and oxygen,of which their organic part is composed,may be obtained either from that source or from the air.

The important distinction which thus exists between these two classes of substances,has given rise to two different views regarding the theory of manures.Basing his views on the presence of the organic elements in the air,Liebig has maintained that it is unnecessary to supply them in the manure,while others,among whom Messrs.

Lawes and Gilbert have taken a prominent position,hold that,as a rule,fertile soils,cultivated in the ordinary manner,contain a sufficient supply of mineral matters for the production of the largest possible crops,but that the quantity of ammonia and nitric acid which the plants are capable of extracting from the air is insufficient,and must be supplemented by manures containing them.

A large number of experiments have been made in support of these views,but the inferences which can be drawn from them are not absolutely conclusive on either side,and it is necessary to consider the matter in a general point of view.

Setting out from the proposition already so frequently referred to,that the plant cannot grow unless it receives a supply of all its elements,it must be obvious that if,to a soil containing a sufficiency of mineral matters to raise a given number of crops,a supply of ammonia be added,its total productive capacity cannot be thus increased; and though it may yield larger crops than it would have done without that substance,this can only be accomplished by a proportionate diminution of their number.

In either case,the same quantity of vegetable matter will be produced,but the time within which it is obtained will be regulated by the supply of ammonia.That substance differs in no respect from any other element of plant-food,and used in this way is to all intents and purposes a special manure,and acts merely by bringing into play those substances which the soil already contains.Its effect may not be apparent until after the lapse of a very long period of time,but it ultimately leads to the exhaustion of the soil.

If,on the other hand,a soil be continuously cropped until it ceases to yield any produce,it is manifest that the exhaustion must in this instance be entirely due to the removal of its available mineral nutriment,because the superincumbent air constantly changed by the winds must continue to afford the same unvarying supply of the organic elements,and the power of supporting vegetation would be restored to it,by adding the necessary inorganic matters.

Hence when a soil,which in its natural condition is capable of yielding a certain amount of vegetable matter,is rendered barren by the removal of the crop,it may be laid down as an incontrovertible position,that its infertility is due to the loss of mineral matters,and that it may be restored to its pristine condition by the use of them,and of them only.But the case is materially altered when we come to consider the course of events in a cultivated soil.

The object of agriculture is to cause the soil,by appropriate treatment,to yield much more than its normal produce,and the question is,how this can be best and most economically effected in practice.According to Liebig,it is attained by adding to the soil a liberal supply of those mineral substances required by the plant,and that it is unnecessary to use any of the organic elements,because they are supplied by the air in sufficient quantity to meet the requirements of the most abundant crops.

Other chemists and vegetable physiologists again hold that though a certain increase may be obtained in this way,a point is soon reached beyond which mineral matters will not cause the plant to absorb more ammonia from the air,although a further increase may be obtained by the addition of nitrogen in that or some other available form.It is admitted on both sides,that all the elements of plant food are equally essential,and the controversy really lies in determining what practically limits the crop producible on any soil.

The point at issue may be put in a clear point of view by considering the course of events on a soil altogether devoid of the elements of plants.If a small quantity of mineral matters be added to such a soil,it immediately becomes capable of supporting a certain amount of vegetation,deriving from the air the organic elements necessary for this purpose,and with every increase of the former,the air will be laid under a larger contribution of the latter,to support the increased growth,and this must proceed until the limit of supply from the atmosphere is reached.

At this point a further supply of mineral matters alone must obviously be incapable of again increasing the crop,and it would thus be absolutely necessary to conjoin them with a proportionate quantity of organic substances.Liebig maintains that this limit is never attained in practice,but that the air affords ammonia and the other organic elements in excess of the requirements of the largest crop,while mineral matters are generally though not invariably present in the soil in insufficient quantity.Messrs.

Lawes and Gilbert,on the other hand,believe that the soil generally contains an excess of mineral matters,and that a manure which is to bring out their full effect must contain ammonia,or some other nitrogenous substance fitted to supplement the deficient supply afforded by the atmosphere.In short,the question at issue is,whether there is or is not a sufficiency of atmospheric food to meet the demands of the largest crop which can practically be produced.An absolutely conclusive reply to this question is by no means easy.

The experiments by which it is to be resolved are complicated by the fact,that all soils capable of supporting anything like a crop,contain not only the mineral,but the organic elements of its food in large and generally in greatly superabundant quantity,and it is impossible satisfactorily to ascertain how much is derived from this source,and how much from the atmosphere.

There are in fact no experiments in which the effects of a purely mineral soil have been ascertained.The important and carefully performed researches of Messrs.Lawes and Gilbert were made upon a soil which had been long under cultivation,and contained decaying vegetable matters in sufficient abundance to supply nitrogen to many successive crops,and it would be most unreasonable to assert that the produce they did obtain by means of mineral manures,drew the whole of its nitrogen from the air.On the contrary,it may be fairly assumed that the soil did yield a certain quantity of its nitrogenous compounds,but to what extent this occurs,it is impossible to determine. This difficulty is encountered more or less in all the other experiments,and precludes absolute conclusions.The same fallacy also besets the arguments of Liebig when he holds that the crop,increased by means of mineral manures alone,must derive the whole of the additional quantity of nitrogen which it contains from the air,as appears to be tacitly assumed throughout the whole discussion.

So far from this being the case,it is just as likely that the mineral matters should cause the plants to take it from the soil,if it is there,as from the atmosphere.Taking a general view of the whole question,it is evident that a certain amount of vegetation may always be produced by means of mineral manures,and the quantity obtained is generally much beyond the normal produce of the soil.

But it is still open to doubt whether the largest possible crop can be thus obtained,although the balance of evidence is against it,and in favour of the addition of ammonia,and other nitrogenous and organic substances,to the soil.

In actual practice manures containing nitrogen are more important,and more extensively applied than any others,and the quantity of that element thus used is very much larger than is generally supposed.Twenty tons of farm-yard manure,a quantity commonly applied,and often exceeded on well cultivated land,contain a sufficiency of organic matters to yield about 2-1/2 cwt.of nitrogen.

A complete rotation,according to the six-course shift,contains almost exactly the same quantity of nitrogen,when we assume average crops throughout the whole,and it is thus made up.

The supply is therefore quite sufficient for the requirements of the crop; and when it is borne in mind that a considerable quantity of ammonia and nitric acid is annually carried down by the rain,and that during a long rotation other substances are very generally used in addition to farm-yard manure,it is obvious that the crop need not depend to any extent upon what it derives from the air.What is true of the nitrogenous matters applies with still greater force to the mineral constituents of the manure.

Twenty tons of farm-yard manure contain 32 cwt.of mineral matters,while the average crops of a six course-shift contain only 1088 lbs.,or less than one-third of this quantity.It is obvious,therefore,that in well manured land there must be a gradual increase of all the constituents of plants,but that of the mineral matters is relatively much greater than that of the nitrogenous.

If therefore from any cause the crop produced on a soil to which farm-yard manure had been applied were greatly to exceed the average,the amount of produce,so far as the soil is concerned,would be limited not by deficiency of mineral,but of nitrogenous food.Hence also when farm-yard manure is liberally applied,there is a gradual accumulation of valuable matters,and a progressive improvement of the productive capacity of the soil. It is far otherwise,however,if a special manure is employed,because in that case the crop is thrown upon the resources of the soil itself for all its constituents except those contained in the substance employed,and by persisting in its exclusive use exhaustion is the inevitable result.It would be wrong,however,to infer from this,that special manures are to be avoided.On the contrary,great benefits are derived from their judicious employment,and the circumstances under which they are admissible may be readily gathered from what has already been said.They are agents which bring into useful activity the dormant resources of the soil,they restore the proper balance between its different constituents,and supply the excessive demand of some particular elements.

Thus,for instance,in a soil containing an abundant supply of mineral matters,a salt of ammonia or nitric acid increases the crop,by promoting the absorption of the substances already present.So likewise a soil on which young cattle and milch cows have been long pastured has its fertility restored by phosphate of lime,because that substance is removed in the bones and milk in relatively much larger proportion than any others.

The choice of a special manure is necessarily dependent on a great variety of circumstances,and is governed partly by the nature of the soil,and partly by that of the crop.It is obvious that cases may occur in which any individual element of the plant may be deficient,and ought to be supplied,but experience has shown that,as a rule,nitrogen and phosphoric acid are the substances which it is most necessary to furnish in this way,and which in all but exceptional cases produce a marked effect on the crop.

The other substances,such as potash,soda,magnesia,etc.,occasionally act beneficially,but the results obtained from them are very uncertain,and frequently entirely negative.It has been commonly asserted that phosphates are specially adapted to root crops,and ammonia or nitrates to the cereals,and this statement is so far true,that the former are used with advantage on the turnip,while the latter act with great benefit on grain crops and more especially on oats and barley.

The effect of the latter,however,is more or less apparent in all crops and on all soils,because it promotes the assimilation of the mineral matters already present.But its peculiar importance lies in the power which it has of promoting the rapid development of the young plant,causing it to send its roots out into the soil,and to spread its leaves into the air,thus enabling it to take from those two sources,abundance of the useful substances existing in them.

But it ought to be distinctly understood,that the statement that particular manures are specially suited to particular crops must be assumed with some reservation,because everything depends upon the nature of the food contained in the soil.It is well known that there are many soils in which ammonia acts more favourably on the turnip than phosphates,and *vice versa*,and the difference is often due to the previous treatment.In many cases in which ammonia when first used proved most beneficial,it now begins to lose its effect,and the reason no doubt is,that by its means the phosphates existing in these soils have been reduced in amount,while the ammonia has accumulated,so that a change in the system of manuring becomes necessary.

A general manure may be used year after year in a perfectly routine manner,but where a special manure is employed,the importance of watching its effects,and altering it as circumstances indicate,cannot be over-estimated.The length of time during which special manures have been extensively used has not been sufficient to bring this prominently before the agriculturist,but its importance must sooner or later force itself upon him,and he will then see the necessity for studying the succession of manures as well as that of crops.

Hitherto we have considered a manure merely as a source from which plants derive their food,but it exercises a scarcely less important action on the chemical and physical properties of the soil.Farm-yard manure,which,as we shall afterwards see,contains a large amount of decomposing vegetable and animal matters,yields a supply of carbonic acid,which operates on the mineral constituents,promotes their further disintegration,and thus liberates their useful elements.

It affects also their physical properties,for it diminishes the tenacity of heavy clays; each straw as it decomposes forming a channel through which the roots of plants,air,and moisture can penetrate more readily than through the stiff clay itself.

On the other hand,it diminishes the porosity of light sandy soils,causes them to retain moisture,and generally makes their texture more suitable to the plant.Special manures probably act to some extent chemically on the soil,but the nature of the changes they produce is as yet imperfectly understood.Superphosphates which are highly acid in all probability act powerfully on the mineral substances,and common salt,which,though of little importance to the plant,occasionally produces very striking effects,appears to exercise some decomposing action on the soil.It is difficult,however,to trace the mode in which they operate on a substance of such complexity as the soil.

Lime,as we shall afterwards see,acts by promoting the decomposition of the vegetable matters on the soil,and possibly some other substances may have a similar effect.In the application of manures to the soil there are several circumstances which must be taken into consideration.It is generally stated that they ought to be distributed as uniformly as possible,but this is not always necessary nor even advisable,and certainly is not acted on in practice.

Much must depend upon the nature both of crop and soil.When the former throws out long and widely penetrating roots,the more uniformly the manure is distributed the better; but if the rootlets are short,it is clearly more advisable that it should be deposited at no great distance from the seed.Practically this is observed in the case of the potato and turnip,which are short rooted,and where the manure is generally deposited close to the seed. But this course is never adopted with the long rooted cereals,the manure being usually applied to the previous crop,so that the repeated ploughings to which the soil is subjected in the interval may distribute what remains as widely and uniformly as possible.

In soils which are either excessively tenacious or light,the accumulation of the manure close to the plants has also the effect of producing an artificial soil in their immediate neighbourhood,containing abundance of plant-food,and having physical properties better fitted for the support of the plant.

On the other hand,when a special manure is used alone,and with the view of promoting the assimilation of substances already existing in the soil,the more uniform its distribution the better,because it is essential that the roots which penetrate through it should find at every point they reach not only the original soil constituents,but also the substances used to supplement their deficiencies.

LIQUID MANURE

This term is applied to the urine of the animals fed on the farm,and to the drainings from the manure-heap,which,in place of being returned to it,are allowed to flow away,and collected in tanks,from which they are distributed by a watering-cart,or according to the method recently introduced in Ayrshire,and since adopted in other places,by pipes laid under-ground in the fields,and through which the manure is either pumped by steam-power,or,where the necessary inclination can be obtained,is distributed by gravitation.That liquid manure must necessarily be valuable,is an inference which maybe at once drawn from the analyses of the urine of different animals already given,and of which it chiefly consists.

In addition to the urine,however,it contains also the soluble organic and mineral matters of the dung,as well as a quantity of solid matters in suspension,among which phosphates are found,and thus it possesses a supply of an element which would be almost entirely deficient if it were composed of urine alone.

The differences are here very remarkable,especially in the quantity of ammonia,which is exceedingly large in the first sample.All of them are particularly rich in potash,and contain but a small proportion of phosphoric acid.The general inference to be deduced from them is,that liquid manure is a most important source of the alkalis and ammonia,and must be peculiarly valuable on soils in which these substances are deficient.

The system of liquid manuring,originally introduced by Mr.Kennedy of Myremill,Ayrshire,and which has since been adopted in some other places,differs from liquid manuring in its *strict* sense,for not only are the drainings of the manure-heap employed,but the whole solid excrements are mixed with water in a tank,and rape-dust and other substances occasionally added,and distributed through the pipes.

It has been abandoned on Mr.Kennedy's farm,but is in use at Tiptree Hall,and on the farm of Mr.Ralston,Lagg,where the fluid is distributed by gravitation.The arrangements employed by Mr.Mechi are identical with those formerly in use at Myremill.The greater part of the stock is kept on boards,and the liquid and solid excrements are collected together in the tank,and largely diluted before distribution.

The quantity of this liquid distributed per acre is about 50,000 gallons,at a cost of 2d.per gallon.As this quantity contains about 39 lbs.of ammonia,it must be nearly equivalent to 2 cwt.of Peruvian guano,which costs,with the expense of spreading,from 28s.to 30s.per acre,while the cost of distributing the liquid exceeds £1: 17s.per acre.

On the other hand,the rapidity with which liquid manure produces its effect must be taken into account.It is on this that its chief value depends,and especially when applied to grass land in early spring,it produces an abundant crop just when turnips and other winter food are exhausted.

Mr.Telfer,Cunning Park,who has used this system for a good many years,has come to the conclusion that it is only in this way that it can be made profitable; and though pipes are laid all over his farm,he has latterly restricted the use of the liquid manure entirely to Italian ryegrass.Its effect on the cereals is much less marked,and it can scarcely be considered as capable of advantageous application to the general operations of the farm.Neither can liquid manure be applied to all soils.It fails entirely on heavy clays,but is peculiarly adapted to light sandy soils; and even barren sand may by its repeated application,be made to yield luxuriant crops.

It is not likely that the system of liquid manuring will extend,except in localities where it is possible to distribute it by gravitation; and even then,it will probably be found most economical to restrict its use to one portion of the farm; and for that purpose,the poorest and most sandy soil ought to be selected.

SEWAGE MANURE

The use of the sewage of towns as a manure is closely connected with that of the liquid manure produced on the farm.Its application must take place in a similar manner,and be governed by the same principles.Although numerous attempts have been made to convert it into a solid form,or to precipitate its valuable matter,none of them have succeeded; nor can it be expected that any plan can be devised for the purpose,because the most important manurial constituents are chiefly soluble,and cannot be converted into an insoluble state,or precipitated from their solution.In its liquid form,however,sewage manure has been employed with the best possible effect in the cultivation of meadows.

The most important instance of its application is in the neighbourhood of Edinburgh,where 325 acres receive the sewage of nearly half the town,and have been converted from barren sand into land which yields from £20 to £30 per acre. It is interesting to notice that this sewage is superior in every respect to the liquid manure used at Tiptree Hall; and the good effects obtained from its application,in the large quantities in which it is used in the Craigentinny meadows,may be well imagined.

It operates,not merely by the substances which it holds in solution,but also by depositing a large quantity of matters carried along in suspension,and is in reality warping with a substance greatly superior to river-mud.

And even,though containing more than half its weight of water and 20 per cent of sand,this substance has considerable value as a manure.The growing evils of the existing system of sewage,and the enormous waste of a manurial matter,which the experience of the Craigentinny meadows has shewn to be productive of the most important effects,has recently directed much attention to the conversion of the contents of our sewers into a useful manure.

Numerous plans for its precipitation and conversion into a solid manure have been proposed,but most of these have shewn an entire ignorance of the fundamental principles of chemistry,and the best only succeed in precipitating a very small proportion of its valuable matters,and leave almost the whole of the ammonia,as well as the greater part of the fixed alkalies,in solution.

Nor is it to be expected that any process will be discovered by which these substances can be precipitated,because solubility is the special characteristic of their compounds,and no means is known by which it is possible to convert them into an insoluble form.If sewage is to be used at all,there seems little doubt that it must be by applying it entire,and in the liquid state.

But here again,the expense of conveying it on to the land becomes an obstacle which it must frequently be impossible to overcome.When it can be conveyed by gravitation,as is the case in the neighbourhood of Edinburgh,it may undoubtedly be used with the utmost advantage,and with the very best economic results.

But when it requires to be carried to a great distance through pipes,and raised to a high level by pumping,all these advantages disappear.If the cost of application amounts to 2d.a gallon,as in Mr.Mechi's case,or even to half that sum,it may be fairly concluded that it cannot be used with any great prospect of large economic results,and that,unless under very exceptional cases,it must be unprofitable.

The chances of success must also greatly depend upon the kind of soil on which it is used.Experience has shewn that its effects are most beneficial on light and deep sandy soils,but that on heavy retentive clays it is without effect,or even absolutely injurious.In clay soils it is important to use every means of getting rid of moisture,and any plan which adds 200 or 300 tons of water to them,only aggravates their natural defects to an extent which more than counterbalances the benefits derived from the manurial matter it contains.

Whatever the ultimate result of the use of town sewage in the liquid form may be,it is unlikely that it will be employed in general agricultural practice.It is more probable that it will be found necessary to set apart a certain breadth of land to be treated by it exclusively.

Many plans have been proposed for conveying it through considerable districts,and selling to the surrounding farmers the quantities which they require,but wherever large sewage-works are established,it will be impossible to depend on a precarious demand,and the promoters of such schemes will be compelled,as part of their speculation,to supply not only the manure,but the land on which it is to be used.

Indeed,the difficulties attending the whole question are so formidable,that even those who are most anxious to see a stop put to the waste of manurial matter must admit that the prospect of a successful economic result is not encouraging.

Nor is it likely that anything will be done until the whole system of managing town refuse is changed,and in place of deluging it with water,some plan can be contrived which,while fulfilling sanatory requirements,shall preserve it in a concentrated form,or convert it into a dry and inodorous substance.

COMPOSITION AND PROPERTIES OF VEGETABLE MANURES

Many vegetable substances have been employed as manures,either alone or as auxiliaries to farm-yard manure.Like that substance,they are general manures,and contain all the constituents of ordinary crops; but,owing to the absence of animal matter,they in general undergo decomposition and fermentation much more slowly,although some of them contain a so largely preponderating proportion of nitrogen,that they may in some respects be compared to the strictly nitrogenous manures.

RAPE-DUST,MUSTARD,COTTON AND CASTOR CAKE

Rape-dust has long been employed as a manure,and the success which has attended its use has led to the introduction of the refuse cake from some other oil seeds,such as those of mustard and castor-oil,which cannot be employed for feeding.Like the seeds of all plants,these substances are rich in nitrogen,and their ash,containing of course all the constituents of the plant,supplies the necessary inorganic elements.

The amount of nitrogen and phosphates,show also that of water and oil,to which reference will be made in a future chapter,in relation to the feeding value of some of them.The detailed composition of their ash may be judged of from that of the seeds from which they are made,and which have been given under that head.

A general similarity may be observed in the composition of all these substances; they are rich in nitrogen,and contain as much of that element as is found in six or seven times their weight of farm-yard manure,and a somewhat similar proportion exists in the amount of phosphates,and probably of their other constituents.

They have all been employed with success,but the most accurate observations have been made with rape-dust,which has been longer and more extensively used than any of the others.It has been employed alone for turnips,or mixed with farm-yard manure,and also as a top-dressing to cereals.But the most marked advantage is derived from it when applied in the latter way on land which has been much exhausted,and its effects are then very striking.

An adequate supply of moisture is essential to the production of its full effects,and hence it often proves a failure in very dry seasons,and on dry soils.It must not be applied in too great abundance,experience having shewn that after a certain point has been reached,an increase in the quantity produces no benefit,and even sometimes positively diminishes the crop.

The other substances of the same class,in all probability,act in the same way,but as their introduction is recent,and their use limited,less is known regarding their effects.

MALT-DUST,BRAN,CHAFF

The value of these substances as manures is chiefly dependent on the nitrogen they contain,though to some extent also on their inorganic constituents.Malt-dust contains about 4·5 per cent,and bran 3·2 per cent of nitrogen.But they are little used as manures,as they can generally be more advantageously employed for feeding.The value of chaff more nearly resembles that of straw.

Straw

Straw is occasionally employed as a manure,and sometimes even as a top-dressing for grass land.It is generally admitted,however,that its application in the dry state,and especially as a top-dressing,is a practice not to be recommended,as it decomposes too slowly in the soil; and it is always desirable to ferment it in the manure heap,so as to facilitate the production of ammonia from its nitrogen.

Still circumstances may occur in which it becomes necessary to employ it in the dry state,and it will generally prove most valuable on heavy soils,which it serves to keep open,and so promotes the access of air,and enables it to act on the soil.

On light sandy soils it generally proves less advantageous,as its tendency of course is to increase the openness of the soil,and render it less able to retain the essential constituents of the plant.The quantity of nitrogen in straw does not exceed 0·2 per cent,and its value is mainly due to its inorganic constituents and to its mechanical effect on the soil.

Saw-dust

Saw-dust has little value as a manure,as it undergoes decomposition with extreme slowness.It is a good *mechanical* addition to heavy soils,and diminishes their tenacity; and though its manurial effects are small,it sooner or later undergoes decomposition,and yields what valuable matters it contains.

The saw-dust of hard wood is to be preferred,both because it contains more valuable matters than that of soft wood,and because the absence of resinous matters permits its more rapid decomposition.It is a useful absorbent of liquid manure,and may be advantageously added to the dung-heap for that purpose.

MANURING WITH FRESH VEGETABLE MATTER

Green Manuring

The term green manuring is applied to the system of sowing some rapidly growing plant,and ploughing it in when it has attained a certain size,and the success attending it,especially on soils poor in organic matters,is very marked.

It is obvious that this mode of manuring can add nothing to the mineral matters contained in the soil,and its utility must therefore be due to the plant gathering organic matters from the air,which,by their decomposition,yield nitrogen and carbonic acid—the former to be directly made use of by subsequent crops,the latter,in all probability,acting also on the soil,and setting free its useful constituents.Hence those plants which obtain the largest quantity of their organic elements from the air ought to be most advantageous for green manuring.

The plants used for this purpose act also as a means of bringing up from the lower parts of the soil the valuable matters which exist in it out of reach of ordinary crops,and mixing them again with the surface part.Many of the plants found most useful for green manuring send down their roots to a considerable depth; and when they are ploughed in,all the substances which they have brought up are of course deposited in the upper few inches of the soil.

Vegetable matter when ploughed in in the fresh state,also decomposes rapidly,and is therefore able immediately to improve the subsequent crop; and as this decomposition takes place in the soil without the loss of ammonia and other valuable matters,which is liable to occur to a greater or less extent when they are fermented on the dung-heap,it will be obvious that in no other mode can equally good results be obtained by its use.

Many plants have been employed as green manure,and different opinions have been expressed as to their relative values.In the selection of any one for the purpose,that should of course be taken which grows most rapidly,and produces within a given time the largest quantity of valuable matters,but no general rule can be given for the selection,as the plant which fulfils those conditions best will differ in different soils and climates.The plants most commonly employed in this country are spurry,white mustard,and turnips.

Rye,clover,buckwheat,white lupins,rape,borage,and some others,have been largely employed abroad.Some of these are obviously unfitted for the climate of the British Islands; and the others,although they have been tried occasionally,do not appear to have been very extensively employed.The turnip is sown broadcast at the end of harvest,and ploughed in after two months.White mustard and spurry are employed in the same way as a preparation for winter wheat,and with the best results.

The latter is sometimes sown as a spring crop in March,ploughed in in May,and another crop sown which is ploughed in in June,and immediately followed by a third.The effect of this treatment is such that the worst sands may be made to bear a remunerative crop of rye.It is not easy to estimate the addition made by green manuring to the valuable matters contained in the soil,but it is probably far from inconsiderable.

A crop of turnips,cultivated on the ordinary agricultural system,after two months' growth,weighs between five and seven tons per acre,and contains nitrogen equivalent to about 48 lbs.of ammonia,and half a ton of organic matters; but nothing is known as to the quantity produced when it is sown broadcast,and is not thinned,although it must materially exceed this.

Neither is it possible to determine the relative proportions derived from the soil and the air,although it is,in all probability,dependent on the resources of the soil itself,—plants grown on a rich soil obtaining their chief supplies from it,while,on poorer soils,a larger proportion is drawn from the atmosphere.

Hence light and sandy soils are most benefited by green manuring,partly on this account,and partly also,no doubt,because the valuable inorganic matters,which are so liable to be washed out of these soils,are accumulated by the plants and retained in them in a state in which they are readily available for the subsequent crop.

Sea-Weed

Sea-weeds have been employed from time immemorial as a manure on the coasts of Scotland and England,in quantities varying from 10 to 20 tons per acre.Their action is necessarily similar to that of green manure ploughed in,as they contain all the ordinary constituents of land plants.The ease with which all sea-weeds pass into a state of putrefaction,adapts them in a peculiar manner to the manurial requirements of a cold and damp climate.The rapidity of their decomposition is such,that when spread on the land they are seen to soften and disappear in a short time.

They form therefore a rapid manure,and their effects are said to be confined to the crop to which they are applied; but this is probably due to the fact,that they are chiefly used in inferior sandy soils,in which any manure is rapidly exhausted.In good soils there is no reason why their effect should not be as lasting as that of farm-yard manure,which,in many particulars,they considerably resemble.

The method of applying sea-weeds most generally in use,is to spread them on the soil,and plough them in after putrefaction has commenced,and it is on the whole the most advantageous.But they are sometimes composted with lime and earth,or mixed with farm-yard manure,and occasionally,also,they are used as a top-dressing to grass land.On some parts of the western coast of Scotland and in the Hebrides,sea-weed is the chief manure.It gives excellent crops of potatoes,but they are said to be of inferior quality,unless marl or shell-sand is employed at the same time.*Leaves* may be used as a manure,simply by ploughing them in,by composting them with lime,or by adding them to the manure heap.

Peat

As a source of organic matter,peat may be used with advantage,especially on soils in which it is naturally deficient.Dry peat of good quality contains about one per cent of nitrogen,and a quantity of ash varying from five to twenty per cent.

These substances,however,become available very slowly,owing to the tardy decay of peat in its natural state; and in order to make it useful,it is necessary to compost it with lime,or to mix it with farm-yard manure,or some readily putrescible substance,so that its decomposition may be accelerated.It may be most advantageously used as an absorbent of liquid manure,and on this account,forms a useful addition to the manure heap.

8484The observations which have been made regarding the use of these 84substances,lead directly to the inference that all vegetable matters possess a certain manurial value,and that they ought to be carefully collected and preserved.In fact,the careful farmer adds everything of the sort to his manure heap,where,by undergoing fermentation along with the manure,their nitrogen becomes immediately available to the plant; while the seeds of weeds are destroyed during the fermentation,and the risk of the land being rendered dirty by their springing up when the manure comes to be used is prevented.

Cereal Seed Treatments

Fungicidal seed treatments help control soil-borne pathogens that cause seed decay,seedling blight and root rot.Control of these diseases may result in better stands,more vigorous seedlings,and increased yields.Protectant fungicides such as captan,maneb,PCNB,thiram,or fludioxonil (Maxim) help control most types of soil-borne pathogens,except for common root rot and take-all.

Protectant fungicides containing captan,maneb,or thiram are sold under various trade names.Fungicidal seed treatments to protect against the soil-borne fungi that cause common root rot and take all will be discussed under barley seed treatment.

Fungicidal seed treatment controls most,but not all,of the seed-borne diseases of small grains.Seed treatment should be used when the seed is contaminated with smut,scab,or black point fungi.Specific recommendations for these diseases are discussed under barley and wheat.No seed treatment is a substitute for good seed.

Barley

Barley has three smuts: covered smut,black semi-loose smut (nigra smut),and loose smut.Covered smut and black semi-loose smut are surface-borne fungal pathogens that infect the emerging seedling.These two smuts can be controlled by various protectant fungicides.Loose smut infects the embryo of the seed before harvest.

Protectant fungicides do not control loose smut—only the systemic fungicides carboxin (Vitavax or Enhance),triadimenol (Baytan) or tebuconazole (Raxil) will control loose smut in barley.These same products also will control the covered and semi-loose smuts of barley.

Covered Smut

Seed of barley varieties susceptible to covered smut should be treated with a protectant fungicide or with carboxin,triadimenol or tebuconazole.At present,all barley varieties recommended for production in North Dakota must be considered susceptible.

Black Semi-loose Smut

Many varieties are susceptible to this smut.It is indistinguishable from loose smut in the field,but it is borne on the seed surface and infects seedlings just as covered smut does.It is not detected by the embryo test,because it does not infect the embryo.It is controlled by seed treatment with a protectant fungicide or with carboxin,triadimenol,or tebuconazole.

Loose Smut

Many of the barley varieties released in the late 1970s and early 1980s were resistant to the races of the loose smut fungus present at that time.In the mid-1980s,a new race of the loose smut fungus was detected in North Dakota.

Seed treatment for loose smut in barley planted for seed production is especially important.Seed samples of susceptible barley varieties should be sent to the North Dakota State Seed Department for an embryo test to determine the percentage of loose smut infection. Losses from loose smut are about equal to the per cent infection; 5 per cent loose smut represents a yield loss of about 4.3 per cent.If the embryo test shows 2 per cent or greater infection,the seed should be treated with carboxin,triadimenol or tebuconazole,or loose smut-free seed should be used for planting.

Common Root Rot

The fungus that causes common root rot is soil-borne and is widespread in North Dakota soils.The common root rot fungus increases under barley or wheat culture,and it often causes mature plant root rot.The fungus also may cause severe root infection in the seedling stage and if splashed to the leaves and head causes spot blotch and black-point infections,respectively.

A number of seed treatment fungicides are now labeled for wheat and barley for suppression of mature plant root rot and also for suppression of seedling blight due to the common root rot fungus.Imazalil is available under several trade names.Triadimenol (Baytan) also is registered for suppression of seedling blight due to common root rot.Tebuconazole (Raxil) is registered for wheat and barley,and difenoconazole (Dividend) is registered only for wheat at the time of publication of this circular. The conditions in which seed treatment for common root rot would be most beneficial are: where continuous wheat or barley is grown,or if short rotations between these susceptible crops are practiced,and in soils where moisture stress is very likely.

Barley Stripe

This fungus disease is rarely seen in North Dakota but occasionally appears in two-row barley if contaminated seed is planted.The disease causes distinct yellow,then brown stripes that run the entire length of the leaves.The stripes join together and the leaves become split and frayed or shredded in appearance.

The fungus is seed-borne in the hull and seed coat.If seed from a suspected disease source is used,seed treatment with a carboxin + thiram or tebuconazole product provides some control of the barley stripe fungus.The systemic fungicide imazalil provides good control of barley stripe.

CORN

Corn seed is especially susceptible to attack by soil-borne pathogens when sown in cold wet soil,when the seed is in poor condition,when it is mechanically injured,or if it has been stored for two years or more.

Seed treatment will protect against seed rot and reduce the danger of seedling blight.Sweet corn is more susceptible to attack than field corn,but both should be treated.Most field corn is already treated when purchased.A number of protectant and systemic fungicides are registered for control of seedling blights and seed rot in corn.

Wheat

Bunt or Stinking Smut

The fungus causing this disease adheres to the seed surface and then infects the emerging wheat seedling.The bunt fungus has a fishy odor and imparts the same odor to flour made from bunted kernels.The price of bunted wheat is discounted for this reason.Bunt is not known to occur in North Dakota at the present time,nor has it occurred in recent years.However,it occurs in other wheat-producing states in the Great Plains.

Growers who purchase seed from out of state should use a seed treatment fungicide,since out-of-state seed could be infested with (carrying) bunt.Growers who have their crop custom combined and then save their own seed should use a fungicide,since the combine could be contaminated with bunt spores or bunt balls from states to the south.Many protectant,systemic,or combination products are available to control bunt.

Loose Smut

Loose smut of wheat infects the embryo,as with loose smut of barley.However,the wheat loose smut pathogen does not infect barley and the barley loose smut pathogen does not infect wheat.No embryo test is available for detection of loose smut in wheat.

Counts of loose smut-infected heads can be made in wheat and durum fields to estimate the per cent infection in the crop.Counts should be made at flowering time,as loose smut heads are hard to detect later.The counts provide information on the per cent infection in the seed that was planted but are not a reliable estimate of the amount of infection in the seed of the harvested crop.

If weather conditions are favorable,*i.e.*cool and wet,infection in the harvested seed could be considerably higher than in the seed that was planted.Carboxin,difenoconazole,tebuconazole,and triadimenol seed treatments control wheat loose smut.

Scab and Black-point

Scab and black-point are fungal diseases that attack both wheat and barley seed.The scab fungus also may infect oats.The scab fungus infects the kernels during flowering if warm,wet conditions prevail.Black-point infection occurs from heading to maturity and also is favored by warm,wet weather.

Scabby seed is shriveled,light in test weight,and often has a chalky white or pink discolouration.When planted,scabby seed has poor germination and poor vigour.The scab fungus does not grow systemically from the seed through the plant to cause subsequent head scab infection.The source of head infection is spores from infected small grain or corn residue.

Black-pointed seed has a black to brown discolouration of the embryo or germ end,a discolouration which may extend around the kernel into the crease.When planted,black-pointed seed also may have low germination; if the seed germinates,the seedling roots are often infected by the fungus.

Protectant seed treatments or protectants in combination with systemics,or systemic products alone have provided significant improvement in stand and vigour.Yield increases also have occurred with treatment of scabby and black-pointed grain.Seed treatment helps assure good stands and seedling vigour; it will not prevent possible scab or black-point infections of the head during the growing season.

Take-all

Take-all is a root disease of barley and wheat caused by a fungus that thrives in very wet soils.Take-all generally is not a serious or common disease in barley or wheat in North Dakota,unless the crop is grown under irrigation.Triadimenol (Baytan),difenoconazole (Dividend) and tebuconazole (Raxil) are registered for suppression of this disease.

PYTHIUM

Damage to wheat from the *Pythium* fungus has not been well documented in North Dakota,but the fungus is common in agricultural soils and generally does more damage where wheat is planted into crop residues and soil temperatures are cool and soil moisture is high.Several seed treatment fungicides are registered for control of *Pythium* spp.in wheat,including difenoconazole + mefenoxam (Dividend XL),mefenoxam (Apron XL),and metalaxyl (Allegiance).

OATS

Smuts on oats have not been reported as a recent problem in North Dakota,although they have been serious in other states in the region.Oat smut can cause severe losses.Distinguishing between covered and loose smuts of oats in the field is not easy. Oat loose smut is more difficult to control with protectant fungicides than oat covered smut,because the loose smut spores are lodged under the hulls where they are difficult to reach.Carboxin and tebuconazole are registered for oat loose smut control.

Legumes

Alfalfa and Small Seeded Legumes

Alfalfa,clover and other small seeded legumes may be treated for seedling blight diseases and damping off.The extent of occurrence of these disease problems in North Dakota has not been documented. Captan or thiram products are registered for seedling blight control,while mefonaxam,metalaxyl,and oxadixyl products are registered for control of *Pythium* damping off and early season infections by *Phythophthora.*

Soybeans

Generally,seed treatment is not required for soybeans.However,it may pay to treat seed if: the germination is below 85 per cent; the seeds are contaminated with fungi; the seeds are badly weathered or injured; or the seed coats are broken.If captan or PCNB treated seeds are to be inoculated with *Rhizobium* bacteria,the inoculant should be applied in-furrow.

Thiram,fludioxonil,metalaxyl and mefenoxam have little or no effect on *Rhizobium* bacteria.Generally,inoculant is not used on soybeans planted on land that was previously cropped to soybeans.

In fields with *Rhizoctonia* problems,a seed treatment containing carboxin,chloroneb,or PCNB should be used.Seed infested with Sclerotinia should be treated with fludioxonil,which will control infection of seedlings by the Sclerotinia fungus.

Dry Edible Beans

Most dry bean seed is treated prior to sale,often with a combination of fungicide,insecticide and bactericide.A fungicide protects against seed- and soil-borne fungi.The bactericide streptomycin controls surface-borne blight bacteria but will not control internally-borne blight bacteria.

Chickpeas

Several seed treatment fungicides are registered for use on chickpeas.Fludioxonil (Maxim) gives broad spectrum protection against soil-borne fungi; mefenoxam and metalaxyl (Allegiance) provide very specific and highly effective protection against *Pythium* seed rot and seedling blight and early season *Phytophthora* root rot.

Chickpea growers in some parts of western North Dakota have experienced severe difficulty with seed rot of untreated seed and should treat seed with mefanoxam or metalaxyl.

The most serious foliar disease of garbanzo beans is *Ascochyta* blight.*Ascochyta* can be seed-borne,usually at low levels.Wet weather may cause rapid spread of *Ascochyta* from a few infection centers.Thiabendazole (TBZ) is registered for use on garbanzo beans for eliminating seed-borne *Ascochyta*.Purchase of western grown seed that has been laboratory checked for *Ascochyta* also is desirable to minimize the danger of losses from this serious disease,but it should also be treated with thiabendazole.

Lentils

Lentils should be treated with captan,fludioxonil (Maxim),mefenoxam (Apron XL) or metalaxyl (Allegiance).Captan and fludioxonil provide broad spectrum protection against *Rhizoctonia* and *Fusarium* seedling blights.Mefenoxam and metalaxyl (Allegiance) provide excellent control against *Pythium* seedling blight,but do not provide protection against *Rhizoctonia* or *Fusarium*.

The two most serious diseases of lentils are anthrac-nose and *Ascochyta* blight.Both can be seed-borne at low levels but seed transmission to developing seedlings has not been deomonstrated for anthracnose.Ascochyta can spread rapidly in wet weather from a few infection centers.

The *Ascochyta* that infects lentils is a different species than the one that infects chickpeas.No fungicide currently registered is effective for elimination of either of these disease fungi on the seed.Purchase of western grown seed that has been laboratory tested for *Ascochyta* is desirable to help minimize the danger of losses.

Flax

Flax seed can be attacked by seed- and soil-borne pathogens,especially when the weather is unfavorable for germination and growth or when the seed coats are cracked and split.Seed treatment with protectant fungicides captan,mancozeb,maneb or thiram will reduce the amount of seed rot and seedling blight.

Yellow-seeded varieties are more susceptible to seed coat damage than are brown-seeded varieties.All varieties currently being grown are brown-seeded varieties except Omega,which is yellow-seeded.

Peas

Common seedling blights of peas can be controlled with captan,fludioxonil,PCNB or thiram.The water mold fungi *Pythium* and *Phytopthora* are common problems on peas and can be controlled with mefenoxam,metalaxyl or oxydixyl treatments.No seed treatment is registered for Aschochyta blight on peas,but data from Manitoba indicates that thiram provides good suppression of seed-borne Ascochyta.

Safflower

Safflower rust is borne on the surface of the seed and produces infections on the hypocotyl of the emerging seedlings.Seed-borne rust can result in poor stands and reduced vigour of the seedlings.Fungicidal seed treatment of safflower with one of several available fungicides is recommended to control seed-borne safflower rust.Carboxin,mancozeb,and thiram are labeled for seed treatment use on safflower.

The winter spores of safflower rust can survive from one season to the next in the soil but will not survive to the second season.Control of safflower rust requires seed treatment with a fungicide and crop rotation.Never plant safflower on land that had safflower the year before.

Sugarbeets

*Pythium,Aphanomyces,*and *Rhizoctonia* are fungi that may cause stand establishment and seedling disease problems in sugarbeets.The *Pythium* fungus occurs in most sugarbeet soils.It causes seed rot,pre-emergence damping off,and post-emergence damping off.

Post-emergence damping off caused by *Pythium* may occur when the seedlings are so tiny that they dry up and blow away within a day or two.Consequently,Pythium-induced seedling death is seldom noticed by the grower.The only thing noticeable may be an unusually poor emergence and stand.

Aphanomyces and *Rhizoctonia* cause death of seedlings at a later growth stage,with plants progressively dying from the two to the eight leaf stage.Later in the season,both fungi may cause a root rot which weakens the plant and also reduces the weight and quality of beet roots.Both diseases are favored by warm soil conditions.*Aphanomyces* is favored by heavy soils with poor drainage,resulting in saturated or puddled soils.*Rhizoctonia* is favored by moist soils.

Most sugarbeet seed is sold treated,but different treatments vary in their effectiveness against these three fungi.Certain fungicides have specific activity: mefenoxam,metalaxyl (Allegiance) and oxadixyl (Anchor) are highly effective against *Pythium;* thiram is moderately effective against *Pythium;* PCNB,chloroneb and fludioxanil (Maxim) are effective against *Rhizoctonia.*

Growers planting in fields with known *Rhizoctonia* problems may wish to request a special or supplemental seed treatment from their seed supplier or else use a planter box overtreatment.Hymexazol (Tachigaren) is effective against *Aphanomyces* and *Pythium.*

Canola

Blackleg is a fungus disease of canola that can cause severe losses.It is seed-borne and also spread by wind-borne as well as rain-splashed spores.The wind-borne spores come from canola crop refuse.Most long distance spread of the blackleg fungus is on seed.Most spread within a field or between fields is from wind-borne spores.The severe (highly virulent) strain of blackleg is common.

Benomyl seed treatment provides excellent control of the seed-borne phase of blackleg.Registration of other seed treatments effective against blackleg may occur in 2000 or 2001.Seed treatment is highly recommended for all canola seed planted in areas that do not yet have the severe strain of blackleg.In areas where the severe strain is prevalent,it may be necessary to plant tolerant varieties of canola.Other seedling blights may be controlled with captan,fludioxanil or thiram.

Sunflower

Downy mildew is a soil-borne disease that can cause severe losses if excessive rains occur shortly after planting.The downy mildew fungus is widespread across North Dakota.It survives many years in the soil and infects emerging or recently emerged seedlings when the soil is saturated.Several new races have occurred in recent years and at present only one hybrid is resistant to all races. The downy mildew fungus has developed resistance to mefenoxam,metalaxyl and oxadixyl.Seed treatment with these products provides poor to fair control of downy mildew,depending on what per cent of the downy mildew population is resistant.No suitable replacement fungicide was available for the 2000 growing season.

BIOLOGICAL CONTROL

Kodiak

Kodiak concentrate contains *Bacillus subtilis* bacteria which colonize the developing root system,suppressing disease organisms such as *Fusarium,Rhizoctonia,Alternaria* and *Aspergillus* that attack root systems.When used with a chemical seed treatment,the combination of chemicals and Kodiak provides protection to the root for a much longer time than with chemicals alone. As the root system develops,the bacteria grow with the roots extending the protection throughout the growing season.As a result of this biological protection,a vigorous root system is established by the plant,which often results in more uniform stands and greater yields.Registered for seed and pod vegtables,soybeans,wheat and barley,and corn plus all other agricultural seeds.

Safety Precautions of diseases

Seed treatment chemicals are pesticides and should be handled as such.Product labels provide information on safe handling and application.Be sure to follow all instructions as listed.

HAZARDS ASSOCIATED WITH HANDLING

Read the label and follow instructions carefully.Over treatment may injure the seed and under treatment will not control the pests in question.The label will also provide information on the toxicity of the active ingredients in the form of code words. The label will provide basic information on first aid,environmental hazards,directions for use,proper storage and disposal of containers,and pesticide and general product information.Use the label and be familiar with it.Material safety data sheets (MSDS) are available from the company for each of the products sold.Consult the MSDS for more detailed information before applying the material.

When handling toxic substances,operating personnel must be provided with protective clothing such as coveralls,cap,protective glasses,rubber apron,rubber boots,rubber gloves,and a respirator designed for use with the substance.Personnel should not inhale the dust or vapour nor permit the material to contact the skin or eyes.

The operator should wash thoroughly with soap and water before eating and smoking.Bathe immediately after work and change all clothing and wash clothing thoroughly with soap and hot water before reuse.In case of contact,immediately remove contaminated clothing and wash skin thoroughly with soap and water.A safety shower should be installed in the immediate vicinity of the treater.

An exhaust system should be installed to remove vapours and dust from the operating area.The exhaust air should discharge into a cyclone or bag-type dust collector.The treater should also be vented and tied into the exhaust system.Seed treaters should be isolated and not operated in the vicinity of other personnel or farm commodities that are to be used for food,feed,or oil.

Special multiwall paper bags or tightly woven bags are recommended for seed that has been treated with toxic substances.Seed should be thoroughly dry before going into the bag,as excessive moisture can cause rapid deterioration of the seed.Store treated seed in a cool,dry place away from food or feed products. Seed treaters should be thoroughly cleaned after use as some of the pesticides are corrosive and others settle out and cause clogging of the equipment.Do not run contaminated water into a stream or public sewer,but discharge into a shallow ground pit.

Hazards Associated with the use of Treated seed

The user of treated seed should read and follow the label carefully.The planter hoppers should be filled outdoors.Do not breath dust or fumes from treated seed or get in eyes or on skin.Wash thoroughly with soap and water after handling treated seed.Use treated seed as seed and do not introduce them into the food or feed channels because serious injury to livestock and humans may occur.Follow instructions on the label for disposal of seed treatment containers and bags containing treated seed.

Do not reuse pesticide containers.Triple rinse and,if possible,recycle containers.Containers can be punctured and disposed of in a sanitary landfill or incineration,or,if allowed by state and local authorities,by burning.If burning,stay out of the smoke.Old seed with low germination should be destroyed by burying at least 18 inches deep in an isolated area away from the water supplies.Treated seed exposed on soil surfaces will be hazardous to birds and other wildlife.

Biological Seed Treatments

Recent research has focused on biological control of certain seed and seedling diseases of crops.The most promising biological control agents tested have been either fungi or bacteria.These biological seed treatments control seed pests by parasitizing the pest organisms,competing for food on the root system,or producing toxic compounds that inhibit pathogen growth.Several biological seed treatments are currently on the market.

Insecticide Seed Treatments

A limited number of insecticides are labeled for use as seed treatments to protect seedlings of various field crops from soil inhabiting insects.

SOIL INSECT PESTS

Soil insect pests are generally more of a problem on corn and soybeans than on small grains and forages.Soil-applied insecticides recommended for control of rootworm larvae also may prevent early stand losses due to wireworms,grubs,seedcorn maggot,and seedcorn beetles.If a soil-applied insecticide is used at planting,then a seed applied insecticide may not be necessary. Refer to the label of the soil-applied insecticide to determine which insects are controlled before purchasing a seed treatment insecticide.Stand losses to these soil insects is generally not significant enough to warrant routine use of soil-applied insecticides in first-year corn. The use of insecticide for first-year corn may change in the future due to a biotype of the western corn rootworm beetle. Benefits from suppression of early-season soil pests may be significant in:

- No-till or reduced-tillage conditions with poor weed control where a high risk exists for a given insect problem such as wireworms and grubs in corn following established sod,
- Continuous corn where early-season pests may cause unacceptable stand losses.

Seedcorn Maggot

Seedcorn maggot or the bean seed maggot are small,yellowish-white,legless fly larvae found feeding on corn seeds.Extensive feeding by these maggots will cause a reduction in stand. In general,seedcorn maggot problems are most likely to occur in situations where:

- High organic matter or decaying vegetation attracts egg-laying adult flies.
- Cool,damp soil conditions delay seed germination and prolong the period vulnerable to maggot attack.

No-till production systems have not been found to increase the likelihood of problems with seed maggots.Several products are formulated for application prior to planting seed on the farm as planter-box or hopper-box treatments.

Seedcorn Beetle

Partially eaten seeds,loss of germination,or stunted seedlings in the presence of small (1/4 to 1/3 inch long) brown ground beetles indicate a seedcorn beetle problem.As with seedcorn maggots,damage is most likely to occur under cool,damp conditions where seed germination and seedling development are delayed.

Wireworms

Wireworms are the larval stage of the "click beetles." The term "wireworm" applies to a complex of species with life cycles requiring one or more years per generation.Wireworm populations affecting corn are most severe in fields following sod or fields having a prolonged grassy weed problem.The larvae pass through a number of life stages,or instars.The earliest stages are very small and white; the latter stages have a characteristic hard-shell appearance and are yellow-brown in colour.

Full-grown larvae range from 1/2 to 1 inch long depending on the species.Wireworm's damage corn by feeding on germinating seeds and young seedlings and may bore into stalks at the soil level.Stand loss may be significant in fields with high populations.In fields where a soil-applied insecticide is not used,especially first year corn fields,application of seed treatment insecticide is strongly recommended.

When to Treat Seed

Corn

Treatment is recommended for all seed corn to prevent or reduce seed decay and seedling blights.The germination and early growth of the seedling is a critical period in the life cycle of corn.Soilborne organisms may invade and kill the embryo before germination,or when invasion occurs later,the seedlings may be destroyed before or after emergence.Seedlings surviving fungal infections at germination are often less vigorous.

Corn seedling diseases are more prevalent in cold,wet soil than when soil temperatures are above 55 degrees F.Therefore,early planted corn needs the added protection of a good seed treatment fungicide.Seedling blights tend to be more severe in fields with poor seed bed preparation such as in no-tillage or reduced-tillage fields.Corn seed is generally treated with a fungicide or fungicide-insecticide combination by the seed producer or seed processor.

Corn smut,leaf blights,stalk and ear rots,and virus diseases are not controlled by seed treatment fungicides.

Small Grains

Seed treatment of small grains is recommended to control smut diseases and to reduce seed decay and seedling blights.The smut diseases include stinking smut (common bunt) of wheat,loose smut of wheat and spelt,and the loose and covered smuts of oats and barley.

All of the wheat varieties planted are susceptible to stinking smut (common bunt).Wheat and oat varieties have resistance to some,but not all,races of loose smut (wheat) and loose and covered smuts (oats).The only sure method of smut control is by using an effective seed treatment fungicide.Spelt is extremely susceptible to loose smut and all seed should be treated with an effective fungicide. Seedling blight phases of head scab (*Gibberella zeae*) and glume blotch (*Stagonospora nodorum*) contribute to poor stands of small grains.Both seed infecting fungi contribute to poor quality seed by causing lightweight,shriveled grain.Since these seed-borne diseases are common,all small grain seeds should be treated to control seedling blight when infected seed are planted.

Soybean

Seed treatment for soybean is recommended in three disease situations:

- When poor quality seed is used for planting,
- When early season control of Phytophthora damping-off is needed,
- When planting early into cool,wet soils,especially in reduced tillage.

The major cause of poor-quality soybean seed is Phomopsis seed rot.Infected seed may be visibly moldy,yet others appear healthy.Purchasing certified seed with the per cent germination,the variety name,and the seed treatment material used listed on the label is one way to ensure quality seed is planted.Germination percentage of bin-run beans is generally unknown unless they are tested.

Seed lots with greater than 80 per cent germination generally do not benefit from seed treatment when planted at recommended seeding rates in warm soil.Those with less than 80 per cent germination may benefit from seed treatments by increased stand and,possibly,increased yield.Decisions to use poor-quality seed should be based on the fact that any seed treatment fungicide effective against seed-borne diseases will not increase germination more than 20 per cent.

If bin-run beans of 50 per cent germination are treated with a fungicide,the grower should not expect over 70 per cent germination in the field.Planting rates can be increased to compensate for seed of low germination.

A number of fungicides are effective in controlling Phomopsis seed rot.These materials have varying degrees of activity against seed and seedling diseases.

Phytophthora damping-off is a serious disease of soybean seedlings in the more heavy,poorly-drained soils.The increase in no-tillage and reduced tillage has increased the incidence of Phytophthora damping off in the state.No-till soils remain wetter longer in the spring and less precipitation is required to saturate them compared with plowed soil during the early stages of crop development.

Extra moisture in the soil pore spaces favors germination of *Phytophthora*.It is recommended that seed treatments be used even when planting *Phytophthora* resistant varieties to ensure good stand establishment.Rhizoctonia seedling blight has caused considerable stand losses over the last decade.Damage from *Rhizoctonia* appears to be a stress related in that factors that adversely affect germination and establishment of seedlings predisposes the plants to infection.Stress from planting in dry soil,herbicide injury reduced tillage fields and shallow planting appears to increase the potential for damage by

Rhizoctonia.Adequate seed bed preparation,soil fertility,and soil moisture for rapid emergence lessens the potential for problems from *Rhizoctonia*.

Sclerotinia stem rot may be introduced into new fields by sclerotia-contaminated seed lots or infested seed.Seed should be well cleaned to remove sclerotia and soybean seed should be treated with a fungicide to eradicate the fungus on infested seed.

Alfalfa and other Small-seeded Forage Legumes

Although a number of fungicides are registered for use as seed treatments of alfalfa,clover,etc.,these crops usually do not respond to seed treatments under field conditions.Some increases in stand may result but these increases have not been reflected in increased yield.The one notable exception to this is the use of fungicide seed treatment on alfalfa for control of seedling damping-off caused by *Phytophthora*.

Significant improvement in stand establishment of alfalfa has been noted in research plots using treated seed.In areas where Phytophthora root rot has caused a serious problem in reducing alfalfa stands or preventing establishment of alfalfa,a variety with resistance to *Phytophthora* that has been treated with a *Phytophthora* specific seed treatment is highly recommended.

Rhizobium Inoculation and Seed Treatments

Rhizobium and Bradyrhizobium are bacteria which can fix nitrogen from the air or collect nitrogen from the soil water solution and incorporate it into compounds that plants can use.The nodules that form on soybean roots are the sites where the bacteria reside.Rhizobia are present in soils,but the commercial strains and some newer strains identified by the USDA are much more efficient at fixing nitrogen.The bacteria may be applied to the seed or placed in the seed furrow at planting.

Insecticide and fungicide seed treatments can have a negative impact on inoculants applied to soybean seed.Several factors affect the impact of seed treatments on seed applied inoculants including: toxicity of material,toxicity of carrier or formulation,length of time inoculant is in contact with seed treatment,and formulation of inoculant.Insecticide seed treatments tend to be the most toxic of all seed treatments to inoculants.The fungicides themselves are not necessarily the problem,but the formulation or carriers may inhibit Rhizobial growth and colonization.

Using fungicides in combination with inoculants can be successful if some precautions are taken.If a specific material is highly toxic to inoculants,then placing the inoculant directly in the furrow and not on the seed is recommended.Treating the seed first and allowing the seed treatment material to dry followed by using humus preparation of inoculant may be successful. Liquid inoculants and fungicides should not be mixed and applied simultaneously.Producers should minimize the time the inoculants are in contact with the seed treatment fungicides prior to planting.

Corn Diseases

Damping-off and seed decay.Germinating corn seed can be attacked by several seed-borne and/or soilborne fungi.Pre-emergence and post-emergence damping-off caused by fungi is most common in poorly drained,cold soils.Seed rots and seedling blights are commonly caused by *Pythium* and *Fusarium* species,and may be caused by *Penicillium* and *Bipolaris* species.All of these fungi can rot seed prior to germination.Infected seedlings typically show a marked softening of stem tissue at the soil line.

Kernels with surface cracks caused by mechanical harvesting are especially susceptible to seed rots caused by soilborne pathogens.Broad spectrum,protectant-type fungicides are highly recommended for both corn and sorghum,especially when early planting in cold,wet soils is attempted.Pythium seedling blight is most significant under these conditions.

Most seed corn companies treat their seed prior to bagging.However,planter-box formulations of these materials are available for use by the grower on untreated field,sweet,and popcorn seeds.

Soybean Diseases

Phytophthora damping-off and root rot is caused by the fungus *Phytophthora sojae*,and is particularly a problem in low,poorly drained,clay soils.However,the disease can be encountered on a variety of soil types if the soil remains wet for several days soon after planting.

Phytophthora can attack soybean plants at any stage of development and stands can be reduced by seed rotting and pre-emergence damping-off.Young plants can be killed soon after emergence.Brown,water-soaked stems and yellow,wilted leaves are the primary symptoms of post-emergence damping-off.

Phytophthora survives in the soil as thick-walled spore,called oospores.Early in the growing season,when the soil remains wet for several days,these spores will germinate,producing a second type of spore called a sporangium.The sporangium then germinates to produce a third type of spore called a zoospore in large numbers.

Zoospores are attracted to chemicals released by soybean roots and swim through a thin film of water until a root is encountered.The zoospores produce a hyphae (a threadlike structure) and infect the soybean root.The fungus grows in the roots and eventually into the plant stem.As the plant dies,oospores are formed by the fungus.

Pythium damping-off can be caused by several species of *Pythium*.Pre-emergence damping-off,seed decays,and root rots of soybean all can be caused by this group of fungi.Pythium rots can occur on soybeans at any stage of plant development but are much more prevalent on seedlings.Diseases caused by *Pythium* can occur over a wide range of temperatures,but are most common during cold periods with high soil moisture.

Seedlings infected with *Pythium* often fail to emerge.Stems of infected seedlings appear water-soaked and translucent above the soil line.Diseased areas later turn brown and appear shrunken; eventually,the stem and smaller roots decay and the seedlings die.

Pythium species survive also as oospores.During periods of high soil moisture,the resting spore germinates and infects seeds or young plants in much the same way as *Phytophthora*.Germinating soybean seeds release a wide variety nutrients or by-products that can stimulate the growth and attract spores of the fungus.Many different seed treatment materials will effectively prevent losses from Pythium seed rot and seedling blights.

Rhizoctonia seedling blight is caused by the fungus *Rhizoctonia solani*,and may occur at any time,but a protected cool period of low soil moisture followed by warming soil temperatures with a brief rainy period favors disease development.Seedlings under stress are more susceptible to Rhizoctonia seedling blight.Example of stress are herbicide injury or desiccation during early seedling growth.

Seedlings and young plants are most often affected.Infected plant stems and older roots appear reddish brown and have sunken lesions.Stems may be girdled just above the soil line; tissue thus damaged may appear cracked or cankered.In dry,windy weather,severely infected plants wilt and die rapidly.

Rhizoctonia solani survives in the soil as sclerotia.The fungal mycelium body itself may also survive in soil or in association with old plant residue.Growth of the fungus in soil depends on soil nutrient supply,pH,moisture,and temperature.

As the fungus grows in soil,it can encounter and infect germinating soybean seeds or young plants.Several seed treatment materials can control the seed rot phase of this disease,however,none are highly effective against the post-emergence or stem rot phase of the disease.

Phomopsis seed rot is a common seed disease and is caused by the seed-borne fungi,*Phomopsis longicolla,Diaporthe phaseolorum* var.*sojae*,and *D.phaseolorum* var.*caulivora*.Infected seeds typically germinate poorly or not at all.

Severely infected seeds appear shriveled,cracked,and elongated,and may be covered with a white,moldy growth.However,seeds may be infected and fail to show any symptoms at all.Infection of seeds is most common when warm,wet weather delays harvest.When moldy seeds are planted,seedling death results in poor stands. Generally,higher soil temperatures favors development of seed rot and seedling blight.Seed treatments can increase the germination rate up to 20 per cent.Seed lots with less than 80 per cent germination should not be used.

Sclerotinia

Recent research demonstrates that Sclerotinia white mold can be introduced into new fields on infested soybean seed.In addition,this fungus,*Sclerotinia sclerotiorum*,forms hard,black,small,irregular-shaped sclerotia both inside and outside of infected soybean plants.

These sclerotia are harvested with the seed or returned to the soil with plant debris at harvest.Seed should be well cleaned to remove sclerotia and treated with appropriate fungicide seed treatment to eradicate the fungus from infested seed.

Wheat and Spelt Diseases

Common bunt or stinking smut is caused by the seed-borne fungus *Tilletia laevis*.The disease causes reduced wheat yields and grain quality.Common bunt was once the most prevalent disease of wheat,however,today the disease is rarely found due to the widespread use of effective seed-treatment fungicides.

Heads of diseased plants may appear stunted and are noticeably thinner and darker in colour than healthy,disease-free heads."Bunt balls" (kernels containing millions of fungal spores) replace the kernels in the flowering heads and spread the glumes much farther apart than normal.Bunt balls approximate the size of normal kernels but tend to be more spherical and off-coloured (typically gray-coloured).When the bunt balls are ruptured at harvest,they release spores into the surrounding air and contaminate the surface of healthy seed.

Bunt fungi survive as resting spores on contaminated seed.When contaminated seeds are planted,bunt spores germinate in the presence of moisture and infect the wheat seedlings.Once established,the fungus grows systemically in the plant.At flowering,entire flowering heads may be infected,and the fungus converts the developing kernels into bunt balls.Healthy seeds are contaminated by the release of the spores at harvest.

Loose smut,caused by the fungus *Ustilago tritici*,has little effect on seed quality.However,it can result in substantial yield losses if left unmanaged.Symptoms are most noticeable between heading and maturity of the crop.Diseased flowering heads are conspicuously darker when compared to healthy,green heads.Also,diseased heads typically emerge earlier.Diseased spikelets may be entirely transformed into dark fungal spore masses.

As the head emerges,floral tissue is torn and the spores are released.Eventually only a spike is left where a normally healthy head should be.Spores of the fungus are dispersed by wind to nearby healthy flowers where they initiate new infections during wet weather.Unlike bunt,the loose smut fungus infects the developing kernels without causing noticeable damage. *Ustilago tritici* survives in infected wheat seed.The fungus becomes active when the seed germinates and grows into the growing point of the developing wheat plant.Fungal growth closely follows the plants' growth.When the flowers form,the fungus again sporulates and starts the cycle again.Use a systemic fungicide seed treatment to eliminate this pathogen from infected seed.

Head scab,caused by *Fusarium* spp.,is frequently a major cause of poor quality wheat seed.Infected seed appear whitish to reddish in colour and are nearly always shrunken and light in weight.Scabby kernels may be dead or the young seedling may be killed by seedling blight after emerging from the soil.Roots of plants killed by seedling blight appear light to reddish brown in colour and may be covered with mold.

If they survive,they generally lack vigour and may produce only a few weak tillers.Planting infected seed early when soil temperatures are above 60 degrees F generally increases loss from seedling blight.The best control of the seed-borne phase of scab is to first clean seed lots to remove all lightweight seed,thus increasing test weights; and second,treat with a fungicide that is effective in controlling seedling blights.

Stagonospora glume blotch,caused by the fungus *Stagonospora nodorum* has increased in the US in recent years probably due to expanded use of nitrogen fertilizers,semidwarf wheat varieties,and reduced tillage.

Symptoms can develop throughout the growing season on all above-ground plant parts.Initial symptoms are small chlorotic flecks usually on lower leaves or those in contact with soil.The glumes of the wheat heads become infected in late spring before flowering.Lesions generally begin at the glume tips and are dark brown in colour.During favorable weather,the fungus penetrates the glumes and infects the seed causing severe shriveling of the grain.

Stagonospora survives on seed,straw,and/or volunteer wheat.With the onset of moist weather,spores are produced and are spread to healthy wheat plants by splashing rain.Glume blotch is most common in relatively warm weather when the temperature is between 70 degrees F and 80 degrees F.

Barley Diseases

True loose smut of barley is caused by the fungus *Ustilago tritici*.The disease cycle is exactly the same as that of loose smut of wheat.Infection occurs when the flowers develop.The seeds become infected internally and must be treated with a systemic fungicide to eliminate the pathogen.

Semiloose smut is caused by the fungus,*Ustilago avenae*.The disease typically causes small losses due to the use of resistant cultivars and fungicide seed treatment.It is most common when soil temperatures are cool and the soil is dry.Distinguishing semi-loose smut from true loose smut in the field is extremely difficult.Diseased heads typically appear earlier than healthy heads.Millions of fungal spores may be formed in diseased heads. *Ustilago avenae* survives as teliospores on the seed surface.At seed germination,the fungus becomes active and infects young seedlings prior to their emergence from soil.As the plants develop,the fungus grows within the growing point of the young plant.At the boot stage,the fungus transforms the kernels and glumes into dark spore masses very similar to those produced by the true loose smut fungus.Spores are readily dispersed by wind or the action of harvesting equipment.Healthy seeds are contaminated upon contact with the spores.

Covered smut caused by the fungus,*Ustilago hordei*,can result in significant yield losses if left untreated.It is a relatively common disease throughout the barley-growing regions of the world.Problems with the disease occur over a wide range of temperatures and are most common under moist soil conditions.

Ustilago hordei survives as thick-walled resting spores on barley seed.Typically,spore germination coincides with seed germination,it is at this time that infection occurs.Once established,the fungus invades the actively growing tissue and keeps pace with growth of the plant.At flowering,the fungus grows through the floral tissue and forms masses of spores in place of healthy seed.Each kernel becomes a spore filled "smut ball" that is enclosed by a thin membrane until plant maturity.At harvest,spores are liberated and may contaminate healthy seed.

Barley stripe,caused by the seed-borne fungus,*Drechslera graminea*,is most common during periods of rain or high humidity.Few seeds are produced on infected plants,and losses generally coincide with disease levels in a particular field.Properly treated seed is the primary means of controlling this disease.Symptoms are first noticeable on the first two or three leaves produced by the plant and on many leaves produced later.

A yellow stripe on the leaf sheath and basal portion of the leaf blade is common.Later,the stripe may extend the entire length of the leaf and the leaf may die.At heading,diseased plants are often stunted and appear a light tan colour in contrast to the larger,green healthy plants.Spikes may be deformed or may fail to emerge.At heading,spores are produced on diseased leaves and spread by wind to nearby heads.Grain produced by diseased plants is typically brown and shriveled.

Drechslera graminea survives entirely as mycelium in the outer layers of the seed.When the seed germinates in moderately moist soil at temperatures below 55 degrees F,the fungus mycelium penetrates and infects the seedling.The fungus grows within plant tissue and occupies the actively growing plant parts.

Net blotch and spot blotch,common diseases of barley,are caused by the fungi,*Drechslera teres* and *Bipolaris sorokineana*,respectively.Net blotch is most common during rainy and humid weather,and early flag leaf infections can result in both reduced grain yield and weight.

A net-like pattern on sheath leaves or flag leaves is the most recognizable symptom of the disease.Initially,small spots or streaks appear on the leaves,followed by an expansion into conspicuous longitudinal and transverse streaks,eventually combining into a network.Brown areas surrounded by greenish-yellow margins are commonly observed on infected leaves.Eventually,entire leaves may wither and die.

Drechslera teres survives in infected seed or on crop residues.Infection of seedlings is most common at 34 degrees F to 60 degrees F.Once infection is established,the fungus produces spores on leaf tissue during periods of high relative humidity.Spores are released and carried by wind to nearby barley plants where new infections can result.Infection of leaves can occur when leaves are wet for periods of 5 to 15 hours within a 46 degrees F to 91 degrees F temperature range.

Spot blotch is most commonly a problem during warm,humid weather.Yield losses of up to 36 per cent have been attributed to this disease when the flag leaf has become severely diseased early in the season and has died prematurely.Seedling blight (*B.sorokineana*) in winter barley is favored by planting early in warm soil.Delayed planting of winter barley in cooler soil helps prevent blight.

Small brown spots surrounded by a yellow halo is the characteristic symptom of the disease.Both leaves and leaf sheaths on plants of all ages can become infected.Eventually,the spots coalesce and cover large areas of the leaves.Minute fungal fruiting structures containing spores can sometimes be observed in the larger spots.Severely diseased leaves may eventually die.

Bipolaris sorokineana survives in seed or infested crop residue.Leaf infection results from airborne spores produced on the seed or on residue.Epidemics of spot blotch can occur when there are prolonged wetness periods (greater than 16 hours) and the temperature is above 70 degrees F.

Note that net blotch is more common in cooler,humid weather,as opposed to spot blotch which prefers warm,humid weather.Most common broad-spectrum seed treatment fungicides are effective in eliminating seed-borne *B.sorokineana* and *D.teres* as well as preventing losses from seedling blight caused by them.

Oat Diseases

Loose smut and covered smut,caused by *Ustilago avenae* and *U.segetum*,respectively.When disease outbreaks are severe,both yield and grain quality can be significantly reduced.However,fungicide seed treatments effectively control both diseases.

Symptoms of these diseases are very similar.Smutted oat panicles emerge from sheaths as olive-brown to black fungal spore masses.Smutted panicles remain more compact than healthy ones,and usually all spikelets on a diseased plant are affected.Smutted plants are often stunted and easily detected in a recently headed field of oats.

The thin membrane enclosing the spore masses of *U.avenae* usually rupture and disintegrate after panicle emergence as opposed to those of *U.segetum*,which can persist longer.

Alfalfa Diseases

Phytophthora damping-off and root rot is caused by the fungus,*Phytophthora megasperma f.sp.medicaginis*.Standing water,poor drainage,and generally wet soil conditions favour disease development.This disease can be lethal to seedlings and established plants. Seedlings may fail to emerge or can be killed after emergence.A conspicuous yellowing of leaves,particularly lower ones,is a characteristic symptom.Infected seedlings often wilt.Seedling tap roots have dark brown to black lesions and may be rotted 2-4 inches below the crown.A yellow discolouration of internal root tissue is often present.Tap roots may appear discoloured.

The life cycle of *Phytophthora* on alfalfa is similar to that on soybean,however,the fungi are completely different and do not attack the other plant host.

Seed treatments are labeled for control of Phytophthora damping off.However,to protect stands beyond the seedling stage,alfalfa varieties with resistance to the fungus should be used.

BACTERIAL RING ROT OF POTATOES

Bacterial ring rot is an important disease of potatoes and is one of the main reasons for rejection of seed potatoes from certification programmes.This disease is particularly serious because it has the potential to spread quickly throughout a farm and may lead to severe losses if left unchecked.Ring rot was originally found in Germany in the late 1800's.The causal bacteria were introduced into the United States in the early 1930's and by 1940 were found throughout the country.

Symptoms

Severe ring rot can result in wilting of leaves and stems along with yellowing and death of leaves.Lower leaves usually wilt first,are slightly rolled at the margins,and are paler green than healthy leaves.As wilting progresses,leaf tissues between veins become yellow.In the later stages of disease,margins of lower leaves die and become brittle,and eventually entire stems yellow and die. Frequently,only one or two stems in a hill will develop symptoms and,in some cases,there are no above-ground symptoms at all.Ring rot derives its name from a characteristic breakdown of the vascular ring within the tuber.This often appears as a creamy-yellow to light-brown,cheesy rot.The symptom is most frequently observed when a diseased tuber is cut crosswise at the stem end. In severe cases,the vascular ring may be separated,and a creamy or cheesy exudate can be forced out from this tissue when the tuber is squeezed.On the outer surface,severely diseased tubers may show slightly sunken,dry,cracked areas.Infected tubers are often invaded by secondary decay organisms which may lead to complete breakdown.

Symptoms of ring rot in the vascular tissue of infected tubers are often less obvious t,appearing as only a broken,sporadically appearing dark line,or as a continuous,yellowish discolouration.Because of this,laboratory tests should always be performed to confirm a diagnosis of ring rot.

Causal Organism

Ring rot is caused by the bacterium Clavibacter michiganense subsp.sepedonicus.Ring-rot bacteria survive between seasons mainly in infected seed tubers.They are also capable of surviving 2-5 years in dried slime on surfaces of crates,bins,burlap sacks,or harvesting and grading machinery,even if exposed to temperatures well below freezing.

Survival is longest under cool,dry conditions.Ring-rot bacteria do not survive in soil in the absence of potato debris,but can survive from season to season in volunteer potato plants.Wounds are necessary for penetration of the bacteria into seed pieces.

The pathogen is easily transmitted from diseased tubers to healthy seed pieces during the seed-cutting process.A knife that cuts one infected tuber can spread these bacteria to the next 20-100 seed pieces.Likewise,the bacteria may be spread during planting,particularly if a picker-type planter is used.

Ring-rot bacteria can be moved in irrigation water and by chewing insects,such as Colorado potato beetles and flea beetles.After the bacteria become established in a plant,they multiply and move throughout the plant via the water-conducting tissues.Fortunately ring-rot bacteria are capable of causing disease only in potato,although they may be able to colonize roots of sugar beets.

- Plant only certified disease-free seed tubers.Certified seed potatoes are produced under regulations mandating zero tolerance for ring rot.Although use of certified seed tubers will not guarantee total freedom from ring rot bacteria,it is the best assurance.
- Discontinue use of any lot of seed tubers in which ring rot is found.Seed lots known to be contaminated with ring-rot bacteria should never be planted.
- Before handling seed tubers,all containers,tools,knives and mechanical cutters,planters,and other equipment should be thoroughly washed with a detergent solution,rinsed,and then sanitized with a disinfectant (for current recommendations,

- If ring rot is confirmed to be present,a thorough cleanup must be undertaken.Dispose of all infected tubers away from potato production areas.Clean all surfaces of storages and equipment to remove all mud,dirt and debris and then wash with a strong detergent in hot water applied by a high-pressure washer.After cleaning,sanitize all storages and equipment with a disinfectant.Do not plant potatoes for two seasons in any field in which ring rot has been found.

POTATO PINK ROT,PYTHIUM LEAK AND SEED-PIECE DECAY

Pink rot and Pythium leak,sometimes collectively called water rot,occur sporadically wherever potatoes are grown.These diseases are a problem of mature tubers at harvest and in storage.They are most serious when warm,wet soil conditions persist during tuber formation and at harvest.

When newly-planted seed pieces are exposed to these conditions,Pythium seed-piece decay also can be severe.Major problems with these diseases are usually associated with excessive rainfall or irrigation either early or late in the season,especially on poorly-drained soils.

Symptoms

Pythium seed-piece decay often results in delayed emergence and poor stands.Infected seed pieces become a soft,watery mass in the soil.Symptoms of pink rot in mature plants include brown or blackened roots or stolons,and in severe cases,leaf chlorosis,stunting,wilting,and even plant death.Tubers develop pink rot mostly through diseased stolons,but occasionally infections occur at buds or lenticels.Decay spreads through infected tubers with the advancing margin of the rot usually sharply defined by a dark line,which may be visible through the skin.

Eyes of infected tubers are often dark brown.Decaying tubers remain intact,but are spongy and odorless.If squeezed,a clear liquid will exude.When infected tubers are cut open,the internal tissues turn salmon pink after a 15-20 minute exposure to air,then later become brownish-black.Pythium leak usually develops through harvest wounds in tuber surfaces and begins as a discoloured,watersoaked area.

As with pink rot,the advancing margin of infection is usually bounded by a dark line.Infected tissues are extremely watery,and appear brown or gray.Severely rotted tubers are of a uniform texture resembling a soft,watery paste.

Causal Organisms

Pink rot is caused by several species of the soilborne fungus Phytophthora (NOT the species that causes late blight) while Pythium leak and seed-piece decay are caused by several species of the closely related soilborne fungus

Pythium.These fungi are widely distributed in both water and soil,and their behaviour is similar.

They survive in soil within decaying plant material,or as resistant spores free in the soil.In warm,moist soil,these fungi produce swimming spores that move in water films.Roots can be infected by Phytophthora at almost any stage of plant growth,but symptoms are more severe on younger roots.

Both groups of fungi infect tubers through wounds,but Phytophthora generally infects tubers before harvest,often through stolons.Infection by Pythium usually occurs through harvest wounds,especially at temperatures above 70 F.Seed pieces can be infected by Pythium as soon as they are planted.

- Select areas with well-drained soils for planting potatoes.
- Use a crop rotation away from potatoes for at least 4 years if pink rot or leak have been severe.This may reduce the amount of fungus surviving in the soil.
- Delay planting for at least 2 weeks after plowing down green vegetation as this may temporarily stimulate populations of Pythium fungi.
- Avoid planting in soils colder than 45 F or warmer than 70 F.
- Avoid harvesting infested fields when soils are especially wet or soil temperatures are below 50 F or above 65 F.Stop irrigation well in advance of harvest.
- Avoid bruising tubers during harvest by adjusting equipment properly,keeping digger chains fully loaded and minimizing drops to 6 inches or less.Do not leave harvested tubers lying on warm,moist soils for any length of time as infection with Pythium may occur quickly.
- Leave low spots in fields unharvested if they have been waterlogged and much rot is present.
- Keep tubers cool and as dry as possible during harvest,loading,transit and storage.
- Grade out infected tubers as much as possible before placing harvested tubers in storage.

  Store lots of harvested tubers containing many infected tubers separately from healthy lots.Good airflow through the pile should be provided to dry out leaky tubers.Lots with significant amounts of disease should be marketed as soon as possible as they will not store well.

SCAB OF POTATO TUBERS

Scab is a disease of potato tubers that results in lowered tuber quality due to scab-like surface lesions.There are no above-ground symptoms.Two forms of scab occur.

Common scab occurs in all production areas and is most severe in soils with a pH above 5.5.Another less common form,called acid scab,is important in acidic soils.

Symptoms

Scab symptoms are quite variable.Usually,roughly circular,raised,tan to brown,corky lesions of varying size develop randomly across tuber surfaces.Sometimes scab develops as a rather superficial layer of corky tissues covering large areas of the tuber surface.This is called russet scab.Pitted scab can also occur where lesions develop up to 1/2 inch deep.

These deep lesions are dark brown to black,and the tissues underneath are often straw-coloured and somewhat translucent.More than one of these lesion types may be present on a single tuber.

Although scab symptoms are usually noticed late in the growing season or at harvest,tubers are susceptible to infection as soon as they are formed.Small brown,water-soaked,circular lesions are visible on tubers within a few weeks after infection.Mature tubers with a well-developed skin are no longer susceptible,but existing lesions will continue to expand as tubers enlarge.Thus disease severity increases throughout the growing season.Scab is most severe when tubers develop under warm,dry soil conditions.Coarse-textured soils that dry out quickly are therefore more conducive to scab than are fine-textured soils. A few other conditions can be confused with scab.White,enlarged lenticles,which frequently occur on potato tubers harvested from wet soil,can be mistaken for scab.Usually this condition will disappear when tubers are dried.Patchy russeting,checking,or cracking of tuber surfaces caused by the fungus Rhizoctonia also may be confused with russet scab.A totally different but uncommon disease called powdery scab,caused by the fungus Spongospora subterranea,causes very similar scab-like symptoms.Laboratory examination may be necessary to identify these diseases.

Causal Organisms

Scab is caused by a group of filamentous bacteria called actinomycetes that occur commonly in soil.In soils with a pH above 5.5,Streptomyces scabies is usually responsible for common scab,and is capable of causing all the types of scab lesions.It is commonly introduced into fields on seed potatoes,and will survive indefinitely on decaying plant debris once the soil is contaminated.Because the organism can survive passage through the digestive tract of animals and be distributed.

Blackleg,Aerial Stem Rot,and Tuber Soft Rot of Potato

Blackleg,aerial stem rot,and tuber soft rot are all similar diseases caused by several types of soft-rot bacteria.Blackleg and tuber soft rot occur wherever potatoes are grown.Aerial stem rot is also widespread,but is most severe under sprinkler-irrigation.

Symptoms

Blackleg begins from a contaminated seed piece,but the symptoms can occur at several stages of plant development.In severe cases,entire seed pieces and developing sprouts may rot in the ground prior to emergence,resulting in a poor stand.Blackleg often develops after plants are well up or even in flower.

In this case,stem bases of diseased plants typically show an inky-black to light-brown decay that originates from the seed piece and can extend up the stem from less than an inch to more than two feet.Leaves of infected plants tend to roll upward at the margins,become yellow,wilt,and often die.Aerial stem rot (also called bacterial stem rot or aerial blackleg) is initiated by soft-rot bacteria from sources external to the seed piece.Stem infection can occur through wounds or through natural openings such as leaf scars.Lesions on diseased stems first appear as irregular brownish to inky-black areas.

These enlarge into a soft,mushy rot that causes entire stems to wilt and die.Potato tubers with soft rot have tissues that are very soft and watery,and have a slightly granular consistency.The diseased tissue is cream- to tan-coloured,and often has a black border separating diseased from healthy areas.

In the early stages,soft-rot decay is generally odorless,but later a foul odor and a stringy or slimy decay usually develops as secondary decay bacteria invade infected tissues.Most internal tuber tissues may be consumed by soft rot organisms,sometimes leaving only a shell of skin remaining in the soil.

Causal Organisms

Blackleg,aerial stem rot,and tuber soft rot are caused by two closely related bacteria,Erwinia carotovora subsp.atroseptica and Erwinia carotovora subsp.carotovora.E.c.carotovora is very common and has an extensive host range,including most fleshy vegetables.It survives readily in soil and surface waters such as rivers,lakes,and even oceans.These bacteria are capable of multiplying and persisting in the root zones of many host and nonhost crop and weed species.In contrast,E.c.atroseptica is associated mostly with potatoes.

These bacteria do not survive well in soil for more than one year,unless they are contained within diseased tubers or other potato plant debris.Blackleg is usually caused by E.c.atroseptica carried on contaminated seed tubers.Most lots of seed tubers are contaminated to some degree,but the bacteria are usually dormant and do not cause disease unless environmental conditions are favourable.In contrast,aerial stem rot is usually caused by E.c.carotovora contained in infested soil or introduced to the crop by irrigation water,wind-blown rain,and insects. Tuber soft rot can be caused by either of these soft-rot bacteria. Moisture and temperature are the two critical factors in initiation and development of soft-rot diseases.High soil temperatures and bruising of seed tubers favour seed-piece decay and pre-emergence blackleg.Blackleg in growing plants is favoured by cool,wet soils at planting followed by high temperatures after emergence.

Dense plant canopies and long periods of leaf wetness favour infection of aerial plant parts.Although tuber soft rot can occur at any temperature above 50 F,disease develops best above 75 F.Oxygen depletion in tubers also favours soft rot.When seed pieces in soil or tubers in storage become covered with a film of water,the tissues rapidly become depleted of oxygen.

This also may be induced by soil flooding or improper drying of washed tubers.Once it starts,tuber soft rot can proceed rapidly in storage."Wet" areas may develop in the piled tubers that flow onto ones below,spreading the bacteria.Heat,coupled with condensation on tuber surfaces,can further adversely affect storage conditions,resulting in accelerated "melt" of the pile.

- Plant only certified,disease-free seed tubers.If possible,use whole (B-size) seed tubers that do not have to be cut.
- When receiving seed tubers in bags,do not stack more than five bags high.With bulk or bagged seed,store at 40-45 F until 2-3 weeks before planting,then warm to 55-60 prior to cutting.
- Clean all equipment used for cutting seed tubers thoroughly and then sanitize with an appropriate disinfectant.
- Treat cut seed pieces with recommended fungicide dressings immediately after cutting.
- Plant treated cut seed pieces immediately if soil temperatures are 55-65 F at planting depth.Seed pieces can be held 1-2 weeks at 55-60 F and 95-99 per cent relative humidity to hasten healing of cut surfaces.Condensation on surfaces of seed pieces must be avoided.

- Do not irrigate fields until plants are well emerged.Avoid using surface water for irrigation.
- During crop growth,monitor irrigation and nitrogen fertility to minimize excessive vine growth that will promote leaf wetness within the plant canopy.
- Harvest tubers only after the vines are completely dead to ensure skin maturity.Low spots in the field should be left unharvested if significant waterlogging has occurred.
- Take all precautions to minimize cuts and bruises when harvesting and handling tubers.
- Hold newly harvested potatoes at 55-60 F with 90-95 per cent relative humidity for the first 1-2 weeks to promote wound healing.After this curing period,lower the temperature of table stock to 38-40 F for long-term storage.Never wash tubers prior to storage.

SEED-PIECE DECAY OF POTATO

Fusarium dry rot is an important postharvest disease of potato tubers that causes significant losses in storage and transit of both seed tubers and those for table consumption.It is also a major cause of seed-piece decay after planting.

Symptoms

Infected tubers usually develop a dry rot,but a moist rot may occur if secondary infections with soft-rot bacteria also are involved.Surfaces of infected tubers are sunken or wrinkled,and rotted tissues appear brown or gray to black.

A white or pink mold is sometimes visible on tuber surfaces.When tubers are cut,internal cavities within rotted tissues may contain white,yellow or pink molds.In storage,blue,black,purple,gray,white,yellow,or pink spore masses may develop in these internal cavities.After low-temperature storage,internal tissues often will become firm and dry or even powdery.

Causal Organisms

Fusarium dry rot is caused by several species of the soilborne fungus Fusarium.These fungi are common in most soils where potatoes are grown and survive as resistant spores free in the soil or within decayed plant tissues.Although some infections may develop on tubers before harvest,most infections occur as the fungus enters tubers through harvest wounds.

Small,brown lesions appear at wound sites 3-4 weeks after harvest and continue to enlarge during storage,taking several months to develop fully.The disease develops fairly rapidly at temperatures above 50 F,but lesions will cease enlarging below 40 F.The fungus is only dormant at these low temperatures,however,and will resume growth when tubers are warmed.Fusarium seed-piece decay is really the same disease.Seed tubers may be infected prior to shipment.Decay during transit or storage often accounts for poor quality seed tubers.When these Fusarium fungi are present on seed pieces or in the soil,poor stands may result,especially if cut surfaces of seed pieces are not properly healed.

Fusarium seed-piece decay begins as reddish-brown to black depressions on cut surfaces.These may expand to cover the entire seed piece and often result in a slimy rot when infection by secondary soft-rot bacteria follows.

- Harvest tubers only after the vines are completely dead to ensure skin maturity.
- Take all precautions when harvesting and handling tubers to minimize cuts and bruises.
- Hold newly harvested potatoes at 55-60 F with 90-95 per cent relative humidity for the first 1-2 weeks to promote wound healing.After this curing period,lower the temperature of table stock to 38-40 F for long-term storage.

- Plant only certified,disease-free seed tubers.If possible,use whole (B-size) seed tubers that do not have to be cut into seed pieces before planting.
- When receiving seed tubers in bags,do not stack more than five bags high.With bulk or bagged seed,store at 40-45 F until 2-3 weeks before planting.Then allow seed potatoes to warm prior to cutting.
- Treat cut seed pieces with recommended fungicide dressings immediately after cutting.
- Plant treated cut seed pieces immediately or store them at 55-60 F and 95-99 per cent relative humidity to hasten healing of cut surfaces.Condensation on surfaces of seed pieces must be avoided.

BACTERIAL SPOT,SPECK,AND CANKER OF TOMATOES

Bacterial spot,bacterial speck,and bacterial canker are widespread diseases of tomato that can cause localized epidemics during warm (spot and canker) or cool (speck),moist conditions.

Bacterial spot can cause moderate to severe defoliation,blossom blight,and lesions on developing fruit.Bacterial speck also causes these symptoms but is usually not as severe as bacterial spot.Bacterial canker causes wilt,vascular discolouration,scorching of leaf margins,and lesions on fruit.

Symptoms

Foliar symptoms of bacterial spot and speck are identical.Small,water-soaked,greasy spots about 1/8 inch in diameter appear on infected leaflets.After a few days,these lesions are often surrounded by yellow halos and the centres dry out and frequently tear.Lesions may coalesce to form large,irregular dead spots.

In mature plants,leaflet infection is most concentrated on fully-expanded and older leaves and some defoliation may occur.Spots may also appear on seedling stems and fruit pedicels.In some cases,blossom blight may occur,causing flower abortion.This is more severe with bacterial spot and may result in a split fruit set which is especially troublesome with determinate cultivars intended for mechanical harvest. Bacterial spot and speck can usually be differentiated by symptoms on immature fruits.

Bacterial spot lesions are small water-soaked spots that become slightly raised and enlarged until they are about 1/4 inch in diameter.The centres of these spots later become irregular,light brown,slightly sunken with a rough,scabby surface.In the early stages of infection,a white halo may surround each lesion at which time it resembles the fruit spot of bacterial canker.Small lesions which have not yet become scabby are often confused with lesions of bacterial speck.

Bacterial speck appears on immature fruit as a black,slightly sunken stippling,eventually causing lesions less than 1/16 inch in diameter.Fruit lesions are not initiated on mature fruit in either disease.Primary or systemic symptoms of bacterial canker (from infections originating in seeds or young seedlings) include stunting,wilting,vascular discolouration,development of open stem cankers,and fruit lesions.When affected stems are split open lengthwise,a thin,reddish-brown discolouration of the vascular tissue is observed,especially at the base of the plant.On young seedlings in the greenhouse,lesions may appear as raised pustules on leaves and stems.

These plants rarely survive the season in the field.Secondary symptoms in the field include leaf "firing" (necrotic marginal leaf tissue adjacent to a thin band of chlorotic tissue) and fruit lesions.Spots on fruit are relatively small (1/32 to 1/16 inch) surrounded by a white halo ("bird's-eye" spots).Canker bacteria may also invade internal fruit tissues,causing a yellow to brown breakdown.

Causal Organisms

Bacterial spot is caused by the bacterium,Xanthomonas campestris pv.vesicatoria,which can be carried as a contaminant on the surface of infested seed and has been found to overwinter in soil associated with plant debris.Bacterial speck is caused by another bacterium,Pseudomonas syringae pv.tomato. This bacterium may also be seedborne and can overwinter on plant debris in soil and on the roots of many perennial plants.Bacterial canker is caused by Clavibacter michiganensis subsp.michiganensis,which,unlike the spot and speck pathogens,has the ability to infect tomato plants systemically.It is seedborne and can survive on infested plant debris in soil.

All three organisms may exist at low populations on leaf surfaces of symptomless plants.At the onset of favourable conditions,these low populations can increase rapidly and bacteria can then enter plants through stomata or small wounds and begin infection.Bacteria can spread rapidly with spattering rain and widespread epidemics may develop.

Penetration of tomato fruit occurs through wounds created by windblown sand,breaking of hairs,or by insect punctures.Optimal conditions for bacterial spot and canker are high moisture,high relative humidity and warm temperatures (75 to 90 degrees F).Bacterial speck is more likely to occur under cool (64 to 75 degrees F),moist conditions.

- Rotate tomatoes with non-solanaceous crops with at least 2 to 3 years between tomato crops.Avoid rotation with peppers,which are also susceptible to bacterial spot.
- Plant only seed from disease-free plants or seed treated to reduce any bacterial populations.Treatments include:
  - Fermentation of tomato pulp and seeds at room temperature for 4–5 days;

- Soaking seeds in 0.6–0.8 per cent acetic acid for 24 hr at 70 degrees F;
- Soaking seeds 5ñ10 hr in 5 per cent hydrochloric acid;
- Hot water treatment of seeds (122 degrees F for 25 minutes); or
- Sodium hypochlorite (bleach) treatment [20–40 minute soak of seeds in 1 per cent sodium hypochlorite (20 per cent bleach)].Some decrease in germination may be expected from these treatments.

- Use only transplants free of disease symptoms.
- Carry out proper sanitation of transplant production greenhouses.Remove all weeds and plant debris,clean all tools with disinfectant solution,and wash hands thoroughly before and after handling plants.Water plants early in the day to reduce the amount of time foliage is wet.Do not handle plants when they are wet.After each crop,clean greenhouse walls,benches,etc.,with hot soapy water,followed by thorough rinsing and treatment with a disinfectant.If possible,close up greenhouse after transplant production is completed to allow natural heating during the summer.Use only new plug trays and pathogen-free planting mixes.Avoid growing peppers and tomatoes in the same greenhouse unless pepper seed has also been treated as in step 2.
- In the field,control irrigation to minimize the time foliage is wet and avoid working among wet plants to minimize spread of disease.
- Applications of mancozeb plus copper soon after transplanting may help retard development and spread of bacterial spot and speck.This practice is not particularly effective for management of bacterial canker.Many tomato processors will not accept tomatoes treated with mancozeb or other EBDC fungicides.Check with your processor before applying one of these fungicides.

BACTERIAL WILT OF CUCURBITS

After vine crops begin to run,gardeners and farmers often notice individual leaves with severe wilt symptoms on sunny days.Within a week or two the condition spreads to entire vines which do not recover from the wilt.This disease,called bacterial wilt,is especially common with cantaloupes and cucumbers.Squash and pumpkins may not wilt as rapidly,but may be dwarfed with extensive blossoming and branching.Watermelons are rarely affected.

Symptoms

Wilting of individual leaves or vines of the plant is the characteristic symptom.One or a few leaves wilt and become dull green.The disease spreads from the leaves downward into the petioles and then the stem until the entire plant wilts and dies.

There are other factors,such as vine borers and soil-borne fungal pathogens,that may cause cucurbits to wilt.Sometimes,if an affected stem is cut off near the ground,the sap may be milky in appearance or sticky and,if touched with the finger,the sap will string up to half an inch.This is a helpful test in diagnosis of bacterial wilt,but cannot be depended upon for positive identification.

Causal Organism

This disease is caused by a bacterium,Erwinia tracheiphila,that overwinters in the bodies of the striped and 12-spotted cucumber beetles.In the spring,the beetles emerge from the ground and feed on young plants,introducing bacteria into the leaves or stems.The bacteria reproduce in the water-conducting vessels,producing gums that interfere with water transport.

The beetles and bacteria are so intimately related that controlling the beetles will control infection by the bacteria.Once infection has occurred,however,no control is possible and wilting plants should be removed,if practical.

Management

The only practical management measure is to use an insecticide when seedlings first emerge to control the black and yellow cucumber beetles.Early infections are most severe,but total control depends on applications continuing at frequent intervals during the growing season.

In some cases,if insect pressure is heavy,it may be necessary to apply an insecticide when plants are just cracking the soil,but have not yet emerged.Management of this disease is completely linked with preventing feeding of cucumber beetles on susceptible hosts.

FUSARIUM AND VERTICILLIUM WILTS

Solanaceous crop plants (tomato,potato,pepper,and egg- plant) may be infected at any age by the fungi that cause Fusarium wilt and Verticillium wilt.The wilt organisms usually enter the plant through young roots and then grow into and up the water conducting vessels of the roots and stem.As the vessels are plugged and collapse,the water supply to the leaves is blocked.With a limited water supply,leaves begin to wilt on sunny days and recover at night.Wilting may first appear in the top of the plant or in the lower leaves.The process may continue until the entire plant is wilted,stunted,or dead.

Tomato and potato plants may recover somewhat but are usually weak,unthrifty,and produce fruit of low quality.Peppers typically collapse rapidly and die.Fusarium and Verticillium wilts are rarely significant in field grown tomatoes due to the widespread incorporation in tomato cultivars of genes for resistance to the pathogen.

However,the resurgent interest in planting "heirloom" tomato varieties which do not carry resistance genes has resulted in increased incidence of Fusarium and Verticillium wilts.Additionally,new races of both pathogens have been identified that are capable of overcoming the resistance in many popular tomato varieties. Verticillium race 2 is now common in tomatoes,but its importance in reducing yield is not known at this time.There is very little genetic resistance available to either disease in pepper or eggplant.

Symptoms

Fusarium Wilt

Fusarium wilt symptoms begin in tomato and potato as slight vein clearing on outer leaflets and drooping of leaf petioles.Later the lower leaves wilt,turn yellow and die and the entire plant may be killed,often before the plant reaches maturity.In many cases a single shoot wilts before the rest of the plant shows symptoms or one side of the plant is affected first.If the main stem is cut,dark,chocolate-brown streaks may be seen running lengthwise through the stem.

This discolouration often extends upward for some distance and is especially evident at the point where the petiole joins the stem.Potato tubers may show browning of the vascular ring as well as browning at the stem end and decay where stolons are attached.In pepper,lower leaves do not begin to wilt until roots and the base of the stem have already started to decay.Wilting of the entire plant soon follows.

Dark brown,sunken,and eventually girdling cankers may be seen at the base of the pepper plant.In eggplant,wilting progresses from lower to upper leaves,followed by collapse of the plant.

Verticillium Wilt

Verticillium wilt symptoms on tomato,potato,and eggplant are similar to those of Fusarium wilt.Often no symptoms are seen until the plant is bearing heavily or a dry period occurs.The bottom leaves become pale,then tips and edges die and leaves finally die and drop off.V-shaped lesions at leaf tips are typical of Verticillium wilt of tomato.Infected plants usually survive the season but are somewhat stunted and both yields and fruits may be small depending on severity of attack.

A light tan discolouration in the stem similar to that caused by Fusarium wilt may be found but is usually confined to lower plant parts.The discolouration is typically lighter in colour than with Fusarium wilt.Symptoms on one side of the plant only are sometimes seen. In potatoes the pathogen may be part of a complex that includes,among others,the root lesion nematode and the bacterial soft rot organism,resulting in premature plant death ("potato early dying disease").

Tubers from Verticillium-infected plants may show light brown vascular discolouration,usually restricted to the stem end.In pepper,the lower leaves wilt,then leaf tips and margins dry and turn brown.Brown streaks in the vascular tissue can be observed well up into the plant,which rapidly collapses and dies.

Causal Organisms

Fusarium wilt in solanaceous crops is caused by several different types of the fungus Fusarium oxysporum.These are: F.oxysporum f.sp.lycopersici (tomato),F.oxysporum f.sp.melongenae (eggplant) and F.oxysporum var.vasinfectum (pepper).Fusarium wilt in potato is caused by a complex of up to four different Fusarium spp.All of the Fusarium wilt pathogens are generally specific to their hosts and are soilborne.They are warm weather organisms,and therefore Fusarium wilts are most serious later in the growing season.

Verticillium wilt is caused by the fungi Verticillium albo-atrum and V.dahliae.These fungi attack a wide range of plant species,including cultivated crops and weeds.They are soilborne in field and greenhouse soils where they can persist for many years.V.albo-atrum is a cool weather organism that grows best when soil temperatures are between 65 and 75 degrees F.V.dahliae is more active between 75 and 83 degrees F.

Although disease is retarded by the higher temperatures that favour Fusarium wilt,visible symptoms may appear to be more severe when high temperatures exist,due to restricted water movement in the plant brought about by damage done to the water conducting vessels earlier in the growing season.

- Because Fusarium and Verticillium fungi are widespread and persist several years in soil,a long crop rotation (4 to 6 years) is necessary to reduce populations of these fungi.Avoid using any solanaceous crop (potato,tomato,pepper,eggplant) in the rotation,and if Verticillium wilt is a problem,also avoid the use of strawberries and raspberries,which are highly susceptible.Rotate with cereals and grasses wherever possible.
- Keep rotational crops weed-free (there are many weeds hosts of Verticillium).
- Whenever practical,remove and destroy infested plant material after harvest.
- Maintain a high level of plant vigour with appropriate fertilization and irrigation,but do not over-irrigate,especially early in the season.
- Plant disease resistant tomato varieties,labeled V (for Verticillium) and F (for Fusarium).These disease resistance designations are usually shown in seed catalogues.Fusarium- or Verticillium-resistant varieties of eggplant,potato,and pepper are generally not available.

- If soils are severely infested,production of solanaceous crops may not be possible unless soil fumigation is an option.

BACTERIAL SPOT OF PEPPER

Bacterial spot of pepper is one of the most destructive diseases of pepper in climates where high temperature and frequent rainfall occur during the growing season.The disease causes spots on leaves and fruit; leaf defoliation; and a reduction in plant growth,fruit yield,and quality.Bacterial spot is also a serious problem on tomato,although not all strains of the pathogen can cause disease on both hosts.

Symptoms

Characteristic bacterial spot symptoms can appear on the leaves,fruits,stem,and petioles.On leaves,symptoms begin as small,yellow-green circular lesions surrounded by a yellowish halo.These spots appear water-soaked under wet conditions.As the lesions mature,a general yellowing extending from the area around the lesions develops on diseased leaves and the centre of the spots become brown to black and sunken.

Tissue in the centre of the lesion often dries and breaks away,giving a "shot-hole" appearance to the leaf.When spots are numerous,they may join together and form irregular discoloured streaks along the veins and leaf margins.Edges and tips of leaves may die,then dry and break away,causing leaves to appear ragged.Severely spotted leaves turn yellow or brown and fall from the plant; young leaves can be distorted.Fruit spots begin as green,circular,slightly raised lesions which eventually become brown or dark,raised,and about 1/8 inch in diameter.Centres of the spots become necrotic,corky,and scab-like.

On stems and petioles,lesions are elongated and blackened,and can kill leaflets.Cotyledons are particularly susceptible to bacterial spot; lesions are initially small,sunken,and silvery.They later become darker in colour.

Causal Organism

Bacterial spot of pepper is caused by a bacterium,Xanthomonas campestris pv.vesicatoria,which can overwinter in crop residue in or on soil,on or in seeds,and on wild host plants.Pepper seeds infested with the pathogen are a major source of inoculum for bacterial spot as well as the major means of long distance spread of the pathogen.The pathogen can survive on dried seeds for up to 10 years.

It is not able to survive free in soil for a long time,but can survive up to 6 months in infected crop debris in soil.The bacterium penetrates leaves through stomata and/or wounds,and fruits through wounds created by wind-driven sand,insect punctures,or mechanical injury.Dissemination of the bacterium occurs between and within fields by water-splashing,aerosols,or during cultivating,hoeing,thinning of direct seeded plants,transplanting,or harvesting.

In the United States,numerous physiological races of the bacterium are known,most of which have been found.

- Use pathogen-free seeds and transplants.
  Use sodium hypochlorite-treated seed to reduce bacterial populations.
- Practice crop rotation with non-host plants such as corn and soybean so that peppers are grown only every 3 to 4 years.However,do not use soybeans in the rotation if white mold (Sclerotinia sclerotiorum) has been a problem.
- Deep plow to bury infected crop debris.
- Avoid working in the field when foliage is wet.
- Eliminate wild host plants such as nightshade and ground cherry in and around field.
- Application of copper-containing pesticides may be helpful for preventing development and spread of bacterial spot.
- None of the currently available pepper varieties are resistant to all known races of the bacterial spot pathogen.However,use of varieties resistant to one or more races of the pathogen may provide some control,depending on the races present.

DOWNY MILDEW OF CRUCIFERS

Downy mildew affects all cultivated plants and weeds in the crucifer family.It can be a serious problem in commercial production of cabbage,broccoli,cauliflower,radish,turnip,mustard,collard,and cruciferous greens.Under favourable conditions,it may cause serious losses in the field or may develop after harvest and cause deterioration of product quality during packing and shipping.

Symptoms

Plants can be infected at any stage of development.In seed beds,cotyledons and primary leaves are invaded resulting in fungal growth visible on the underside of the leaf.Later a slight yellowing develops opposite the fungal growth on the upper side of the leaf.

The young leaf or cotyledon,when yellow,may drop off.Older leaves usually persist and infected areas gradually enlarge,turn bright yellow,then become tan and papery.Rarely the affected leaf may develop hundreds of minute darkened specks.Under cool,moist conditions,a white mildew growth can be seen on the underside of infected leaf lesions. Symptoms may appear on other plant parts as well.

The fleshy roots of turnips and radishes may develop an internal,irregularly shaped discolouration extending from the crown downward.The flesh may be brown or black or show a form of net necrosis.In advanced stages,the skin becomes roughened by minute cracks and the root may split open.In radish these symptoms may be confused with those caused by Rhizoctonia.

Causal Organism

The fungus causing downy mildew in crucifers,Peronospora parasitica,overwinters in roots or in decaying portions of diseased plants.Thick-walled resting spores may form in stems,cotyledons,and other fleshy parts of infected host plants.On growing plants,the fungus produces large numbers of spores that are blown about by wind and splashed by rain.

Moisture and temperature are important in the spread and reproduction of this fungus.High relative humidity during cool or warm,but not hot,periods promotes its growth and sporulation.Presence of a water film on the foliage from fog,drizzling rain,or dew allows spores to germinate,infect,and produce more spores on a susceptible host in as few as 4 days.

- Use a crop rotation plan that excludes production of any type of cruciferous crop for at least 2 out of every 3 years.
- Practice sanitary measures such as the use of clean seed beds away from other crucifer production and the destruction of cruciferous weeds.
- Use a planting site and plant spacing pattern that expose plants to full sun throughout the day.
- If severe disease pressure is expected,apply a registered fungicide weekly beginning soon after emergence.
- Disease resistant cultivars are not available for most cruciferous crops.However,some hybrid cultivars of broccoli are resistant or tolerant to downy mildew.

BLACK ROT OF CRUCIFERS

Black rot is the most serious disease of crucifer crops world wide when environmental conditions (relatively high temperature and humidity) are favourable.The disease affects primarily aboveground parts of plants at any stage of growth and causes high yield and quality losses,especially in tropical and subtropical regions during the rainy season.

All vegetables in the crucifer family,including broccoli,Brussels sprouts,cabbage,cauliflower,Chinese cabbage,kale,mustard,radish,rutabaga,and turnip,are susceptible to black rot.Many cruciferous weeds such as Shepherd's Purse,wild mustard,and yellow rocket may also be hosts of this pathogen.

The characteristic black rot symptom on most cultivated crucifer plants is the appearance of yellow,V-shaped lesions along the margins of leaves.The point of the V-shaped lesion is directed toward a vein.When lesions enlarge,wilted tissue expands toward the base of leaves.Eventually the diseased areas become necrotic and the veins turn black or brown.The infection may move down the vascular tissue of petioles and then spread up and down the stems.When stems and petioles of an infected plant are cut crosswise or lengthwise,the black-brown vascular tissue with yellowish bacterial slime is observed.These symptoms may be confused with Fusarium yellows,except that Fusarium causes brown vein discolouration without bacterial slime.

Moreover,symptoms of black rot may vary according to age of host,host genus,species,and cultivar and even environmental conditions.Symptoms on cauliflower may appear as numerous black or brown specks,scratched leaf margins,black veins,and discoloured curds.Many cruciferous weed species do not exhibit any of these characteristic symptoms even when infected.

Symptoms

Causal Organism

Black rot of crucifer is caused by a bacterium,Xanthomonas campestris pv.campestris (Xcc).The bacteria can overwinter in plant debris,in and on seeds from diseased plants,and in and on weeds.The pathogen may survive in diseased crop residue buried in soil for up to 2 years,but not more than 60 days free in soil.

The major source of these bacteria is infected seeds,which enable long-distance spread of the disease.The pathogen is spread within and between fields by splashing water,wind,insects,machinery,and irrigation or drainage waters.

The bacteria infect the cotyledons and young leaves through natural plant openings (stomata,hydathodes) or wounds and then migrate between cells until they reach the xylem tissue where they spread throughout the plant.Free moisture is required for infection by the pathogen.After infection,symptoms may appear on plants within 7 to 14 days under optimum conditions (25 to 30 degrees C).

Effective management of black rot of crucifers depends on the application of the following practices in combination:

- Use black rot-tested,disease-free seed grown in an arid production area.
  - If source of the seeds is unknown,or infested seedlots must be used,treat seed with hot water to eradicate pathogenic bacteria.Cabbage,broccoli,and Brussels sprouts can be treated at 50 degrees C for 25 minutes,while seeds of cauliflower,kale,turnip,and rutabaga are treated for 15 minutes. However,this treatment may reduce the viability of seed. Therefore,some other chemical seed treatments, including, sodium hypochlorite, hydrogen peroxide, and hot acidified cupric acetate or zinc sulfate can be applied to eliminate the bacteria from crucifer plant seeds.

- Use certified disease-free transplants.
- Practice crop rotation where crucifers are grown only every 3 to 4 years to eliminate the inoculum sources from diseased crop debris in the soil.
- Good sanitation practices should be performed to prevent disease spread.
  - Eliminate all volunteer crucifer plants from previous crops and alternative wild host plants within and around the field.
  - Do not apply manure that may contain crucifer residues.
  - Do not use sprinkler irrigation.
  - Avoid working in the field when plants are wet.
  - Do not allow machinery and equipment movement from infested areas to non-infested fields.
  - Deep plow to bury all crucifer residues after harvest.
- Application of fixed copper pesticides in the field may help to reduce spread of the disease.
- A few black rot-resistant cultivars of cabbage and other crucifers are commercially available.These resistant cultivars should be used in crucifer growing regions where black rot is a common problem.

4

Home Vegetable Gardens and Disease

VEGETABLE GARDENING

Vegetable gardening is a rewarding and enjoyable hobby that provides a supply of fresh produce in season and that also can help to reduce a family's food costs.Vegetable crops are threatened by various disease,insect and weed problems that may reduce the yield and quality of the produce and even destroy the plants.Diseases and pests can damage vegetables from the time seeds are planted until after the crops are harvested,and gardeners who fail to follow good growing practices that minimize damage from diseases and pests in their gardens will suffer many disappointments.

The impact of plant diseases on vegetables often appears less direct than that of insects and weeds.Insects usually can be seen with the naked eye,and their activities quickly noted,while weeds are conspicuous,competing directly with vegetable plants for space,nutrients and moisture.Diseases,on the other hand,may make their presence known only when plant growth begins to slow down or when productivity is less than expected.No causal agent may be visible,only symptoms indicating that something is wrong.

Based on general cause,there are two kinds of diseases: biotic and abiotic.Biotic diseases are caused by an identifiable microorganism or infectious agent,for example,bacteria,fungi and viruses.Abiotic diseases result from unfavourable growing conditions or environmental stresses,such as extreme temperatures,too much or too little water,and nutrient imbalances.Both kinds of diseases exist widely in vegetable gardens across Canada.

Plant-feeding insects,mites and nematodes become pests if they injure crop plants sufficiently to interfere with the capacity of the plant to produce food.Some species of insects also transmit plant pathogens.Aphids,for example,transmit several viruses,and cucumber beetles can spread bacterial wilt and squash mosaic virus.The presence of some plant-feeding pests can be tolerated,but control may be necessary if the pests begin to cause significant damage.Home gardeners usually tolerate more pest injury,especially cosmetic injury,to their vegetables than do commercial producers.Many pest species are kept at a low level by natural enemies,such as predators and parasites,and additional control measures may not be required.It is important to realize that the population build-up of natural parasites and predators lags behind that of the pests,so some damage usually is evident.Special practices may have to be employed to keep the pest population at a level where damage will not be serious.

The most effective pest control strategy for the home gardener utilizes a combination of practices; these include planting resistant cultivars,providing optimum soil conditions,fertility and moisture supply,selecting pest-free seed and transplants,employing crop rotations,monitoring carefully for pests,destroying plant residue,cultivating,mulching,and using pesticides only when necessary.Such practices often keep the populations of garden pests at tolerable levels.The role of these and other techniques in helping to minimize disease,insect and weed problems in home vegetable gardens.

MONITORING

Traps and lures: Baits and lures using light,colour,sex attractants (pheromones) or food to draw a pest to a trap may be effective for some pests.Wireworms can be trapped by burying pieces of potato or carrot in the soil,then checking the bait every few days for the insect.Infested baits can be collected and destroyed.Slugs can be drawn to reservoir traps of beer or to a mixture of molasses,water and yeast. Boards laid out in the garden can be used to trap slugs because they look for a cool,damp spot to hide during the day.These pests can then be collected and destroyed daily.Earwigs can be trapped in wooden traps or in containers filled with water containing a small amount of detergent.Growing an insect pest's favorite plant as a trap crop to lure it from the garden also works.However,the pests must be controlled on the trap crop or they will migrate back to the garden.

CULTURAL PRACTICES

Healthy seeds and transplants: Many disease-causing organisms are capable of living in and on vegetable seeds; therefore,it generally is unwise for a person to save seed produced in the home garden.Rather,it should be purchased from dealers who have a reputation for producing or selling high quality,disease-free seed.When growing transplants,gardeners should use high quality seed and a pasteurized growth medium.If transplants or rootstocks are purchased from a supplier,they should be carefully checked for signs of pests and diseases,and any found to be infested should be destroyed.

Crop rotation: Moving the preferred host plant of an insect or disease organism hampers that pest's ability to feed and reproduce.Many disease-causing organisms do not survive long in the soil in which a different crop is planted.

Exceptions include fungi that cause such diseases as fusarium wilt on cabbage,potato or tomato,and clubroot of cruciferous crops; once the soil is infested with these pathogens,it can remain so for several years.Many insects and mites hibernate or lay their overwintering eggs in the soil near their preferred hosts.Moving the garden to a different location,or even switching the host crops from one to another part of the garden,has the effect that,upon emergence in the spring,the pest may find a non-host plant on which it is unable to feed and reproduce.This practice also may help to control weeds,especially if part of the garden is summer fallowed for at least one growing season.Crop rotation has the added advantage of allowing the soil to rejuvenate itself.Cabbage,turnip and potato use large amounts of nitrogen and should be preceded with legumes,for example,pea or bean,followed by fallow and the incorporation of compost or manure.

It is advisable to rotate among vegetable families because different members of a plant family often are susceptible to many of the same pest problems.For example,tomato,potato,eggplant and pepper are in the potato family and share many common diseases and insect pests,as do members of the cucurbit family (cucumber,melon and squash) and the crucifer family (cabbage,broccoli,Brussels sprouts,cauliflower,kohlrabi,turnip and radish).It is desirable to leave as much time as possible between related crops; three to six years is ideal.If rotation is not practical,gardeners should at least alternate the cultivars of vegetables that they are growing,choosing pest-resistant ones whenever possible.

Manual and mechanical methods: Hand-picking large,slow-moving pests,such as caterpillars,Colorado potato beetles and slugs,and dropping them into a container of soapy water or a 5% isopropyl (rubbing) alcohol solution is effective.Shaking the plants or hosing them off with a spray of water dislodges some insects and mites.Barriers are meant to keep a pest away from the crop that it may harm.

Some examples are floating row covers,copper strips,abrasive materials and mulches.Used cans or frozenjuice containers opened at both ends and placed around transplants will protect them from subterranean cutworms.Copper strips repel slugs because their slimy coating interacts chemically with the copper.

Applying bands of abrasive materials,such as diatomaceous earth,stone dust or crushed egg shells,around the perimeter or between rows in a garden helps to deter slugs and crawling insects.Diatomaceous earth is an insecticide made up of the ground shells of tiny sea creatures called diatoms.The silica daggers pierce the skin of the insect,causing dehydration.Soap solutions smother the pests to which they are applied.Tar paper collars placed on the soil around individual crucifer seedlings will prevent root maggot flies from depositing their eggs at the base of the plants. Mulches provide an effective means of weed control in vegetable gardens.Organic materials,such as weed-free straw or grass clippings 7 to 10 cm thick,will inhibit germination of weed seeds brought to the surface by cultivation,retard development of weed seedlings,conserve moisture,and help to maintain a uniform soil temperature.After the crop is removed,the mulch can be incorporated into the soil to enhance its organic matter content.Plastic mulches increase the soil temperature and also conserve moisture.Black plastic mulches,with openings for the crop,will prevent weed growth,except at the opening.

Clear and white plastic mulches,however,permit weed growth,as some light can penetrate through the plastic.Plastic mulches are best suited for warm-season vegetables.Mulches also may prevent soil from being splashed onto plants,thereby keeping the produce clean and reducing the risk of spreading some diseases.Weed mats made of woven fabric allow water and air to reach the soil but screen out light and act as a barrier to emerging weeds,thereby providing effective,long-term control.

Removal of infested plant material: Many fungal diseases can build up and spread rapidly through the production of millions of spores.Prompt removal or destruction of diseased plan residues retards the spread of disease-causing organisms.This is an effective practice with such diseases as gray mold,powdery mildew,and various leaf spots and fruit rots.Removing diseased plant material from the garden at the end of the season eliminates an important source of inoculum for the following spring.

The remaining residues should be rototilled or spaded into the soil to destroy disease organisms and to expose overwintering pests to the elements and to predators.Composting diseased plant material is not recommended because it may not destroy all of the disease organisms,even if the compost heats properly and is turned frequently.In some cases,leaving the remains of insect-infested plants in the garden may help to increase the population of beneficial parasites.

Good growing practices: When watering vegetable gardens,allow the soil surface to dry before another watering.A void frequent,light waterings as they tend to promote disease development and favour the germination of weed seeds.Watering with soaker hoses or in-ground furrows may reduce disease incidence because the foliage stays dry.When using overhead sprinklers,gardeners should be sure that the water is applied during the late morning or early afternoon.

This allows for rapid drying of the foliage.Prolonged leaf wetness favours the development of most foliar diseases.Wide row-spacing ensures rapid leaf drying and reduces the spread of certain pests and diseases by direct contact.However,wide spacing also reduces shading of the soil,thus enhancing the loss of soil moisture and favoring the growth of weeds.Keeping plants as healthy as possible is an important strategy in managing insect and disease problems.Planting into a warm,moist,well-prepared,well-drained seedbed,maintaining a high organic matter content,applying fertilizers correctly,and not cultivating when either the plants or the soil is wet,are recommended practices.Monitoring the crop through the growing season may help to keep pest problems small by facilitating prompt control.Accurate record keeping is suggested to ensure success from year to year.

Companion planting: Many insects prefer to feed on plants belonging to specific families and reject others.For example,the imported cabbageworm feeds on cabbage,radish,kohlrabi and other cole crops; therefore,interplanting unrelated plants,such as onion,with cole crops will help to deter feeding and limit damage by this pest.Furthermore,garlic,onion and other aromatic plants interplanted among other garden crops are believed to repel certain insect pests.Alternating vegetable cultivars or species also reduces the chances of disease spread.Where root-knot and root-lesion nematodes are a problem,interplanting with French marigold or African marigold may help to reduce the populations of these pests.

Altered planting times: Seeding dates of vegetables can be altered to allow the plants to grow at times when they are less likely to be injured by certain pests.This requires knowing the life cycles of the common pest species in the area.In some cases,it may be feasible to plant certain vegetables before or after the pests have passed through their most active feeding period.For example,in Ontario,onion maggot eggs are laid near onion plants in May,and maggots soon attack plants and may cause them to die.However,onion sets planted after June 1 will escape most first-generation maggots.

Resistant cultivars: Some vegetable cultivars are resistant or tolerant to certain pests and diseases,and growing resistant plants is an excellent way to decrease the risk of damage.Certain disease-causing fungi can survive for many years in the soil and may not be managed by routine cultural practices; therefore,planting resistant cultivars is the best means of controlling them.Packages of tomato seed marked VFN produce plants that are resistant to the *Verticillium* and *Fusarium* fungi that cause wilt diseases,and to root-knot nematodes that may attack and cause galls on the roots.Other diseases best controlled by planting resistant cultivars include cabbage yellows,potato scab and cucumber mosaic.Certain vegetable cultivars also are less vulnerable to insect damage than others.For example,Red Pontiac potato is more resistant than other potato cultivars to tuber tlea beetle,and Champion radish seems to be somewhat resistant to the crucifer flea beetle.

Biological control: Beneficial insects can be attracted to the garden by planting some of their favorite nectar- and pollen-producing plants in the garden or nearby.Members of the carrot (dill,caraway,fennel and parsley),mint (catnip,hyssop and lemon balm),and daisy (yarrow and coneflowers) families are known to attract such insects.Providing a source of water and shelter also helps to keep beneficial insects around.Birds and toads also aid in the control of insect pests. Care must be taken when applying pesticides because beneficial insects can be killed as well as the target pest.The bacterium *Bacillus thuringiensis,* or Bt,is a pathogen of certain insects and must be ingested by the larva of the insect to be effective.Only butterfly and moth larvae are susceptible to this product,although a strain of Bt has recently been developed that will kill the Colorado potato beetle.Various formulations of this microbial pestcontrol product are available for controlling larvae of certain moths and buttertlies on home garden vegetables.This and other treatments must be applied when the pest is most susceptible.

Chemical control: *Seed treatment* Planting fungicideor hot-water-treated seed helps to ensure good stands and may avoid the need to replant.Seed treatments kill diseasecausing organisms on the seed and help to protect the vulnerable seed and young seedlings from certain soil-borne disease organisms.

Treated seed is available to the home gardener and is marked 'treated' on the package.Fungicide-treated seed is usually red or some other easily identifiable colour.If untreated seed is used,it should be certified disease-free or be hot-water treated.

Seed can be treated by the gardener,using recommended chemicals according to the manufacturer's directions.Small packets of seed can be treated by tearing off one corner and putting about twice as much chemical in the packet as can be picked up on the first centimetre of the flat end of a toothpick.The packet should be shaken until the seed is thinly coated.*Foliar treatment* Most foliar diseases can be controlled by spraying or dusting plants with an effective fungicide as a preventative treatment.Protectant fungicides work on the plant surface to protect against infection,but they cannot cure established infections.

If a considerable amount of disease is present,it is usually too late to apply a fungicide treatment,except to protect newly emerging leaves.Fungicides should be applied at 7- to 10-day intervals,or as directed by the manufacturer,with reapplication after rain or watering has washed the material away.A thorough covering of the plants is necessary for proper disease prevention.Early detection,timely fungicide applications and removal of diseased leaves are necessary for effective control of many foliar fungal diseases.Gardeners can choose from a wide variety of organic and inorganic fungicides for use in combatting vegetable diseases in home gardens.

The home gardener also has a number of insecticides and miticides to choose from.Most botanical insecticides are derived from plant parts and will break down quickly into harmless substances.Botanical insecticides include pyrethrin and rotenone.Insecticidal soaps are effective against aphids,but they also are non-selective and may upset the natural balance of beneficial insects and their prey. Chemical control of weeds usually is not practical for small home gardens but can be employed for larger gardens and for market gardens.If the decision is made to use a herbicide,extreme care should be taken to avoid accidental drift onto non-target species.In addition,residual soil herbicides should not be used where residues may carryover and affect succeeding crops.

Growers who apply pesticides in their home gardens should use only products that are registered and recommended for the crops being grown and for the specific diseases or pests that are prevalent in the area.It is very important to carefully follow the manufacturer's directions to ensure maximum effectiveness of a pesticide and to minimize potential adverse effects,such as poor control,crop injury and unacceptable chemical residues on the edible produce.Spot treatment is preferable to a general application because it lessens the risk of harming beneficial organisms,humans and pets.

CAUSE OF DISEASE

Light is essential to the life of green plants.Only in very exceptional cases,such as sycamore cotyledons,do tissues turn green in the absence of light.Moreover light is the source of energy whereby the green colouring matter (chlorophyll) is enabled to transform carbon dioxide and water into sugars and starch.The quality (colour) and intensity of light affect the growth of leaves and flowers and the process of transpiration. Moreover,the daily duration of light is of great importance in determining whether a plant will flower or not (photo-periodism).Some plants flower only during long days; others,which normally flower only in the short days of autumn,can be induced to do so in summer if they are artificially darkened for the requisite number of hours a day.

LACK OF SUN LIGHT

Lack of light generally induces etiolation (blanching).In extreme cases there is a yellow colour,long fragile stalk and small leaves; weak light induces straggling soft growth with few leaves,light green in colour.

Too dense a seeding in seedbeds and seed-boxes has a similar effect in producing weak,semi-etiolated plants particularly liable to damping off.

Injuries due to excessive sunlight are seldom seen and in practice it is difficult to differentiate between injury due to too intense a light and that due to the accompanying heat.Sun scald of greenhouse tomatoes in bright sunny weather is almost certainly due to the latter.

Too intense a light is,however,known to have a destructive effect on chlorophyll and it may be for this reason that plants suddenly moved into full sunlight make red pigment (anthocyanin) in their leaves.Such red colouring is regarded as abnormal in cacti and begonias.Aspidistras,camellias,*Cissits antarctica,* etc.,do not flourish in strong light as their chlorophyll is gradually destroyed by it. Tropical species of *Selaginella* with their delicate leaves cannot endure full sunlight,and many ferns require moderate shade.Camellias in strong light often develop yellow spots on their leaves.

Transpiration

Transpiration or evapouration of water from the leaves and other above-ground parts is an essential function in land plants.Not only is it one of the factors which brings about the ascent of sap but,owing to the large consumption of latent heat in changing liquid water into water vapour,transpiration helps to cool the plant.In hot sunny weather there is intense radiation from the surface of bare soil which scorches the moisture still more out of herbaceous plants.

The effect varies greatly between different plants.Potato leaves wilt helplessly,but beetroots,owing to their great ability for obtaining during the night the moisture they require from the soil and dew,have their leaves as erect as ever next morning.Sunflowers transpire especially through their lower leaves which wilt first; begonias,on the other hand,mainly through their flowers.There is little restriction to evapouration from the surface of most petals and it is a common experience that large showy flowers like chrysanthemums and roses wither very readily when cut.

In extreme drought,plants,and even deciduous trees,shed their leaves.A second crop of leaves is then produced when the drought is over.Similarly in very dry summers apple trees shed a large proportion of their young green apples.The dry air of living rooms tends to production of " long-legged " pot plants with only a few leaves near the top of their stems.Out of doors a dry wind may shrivel the edges of leaves,especially of the newly opened leaves of pear,apple and red currant in the spring.Such marginal scorching becomes more pronounced the further one gets from shelter.Marginal scorch of fruit-tree leaves is often,however,a sign of potassium deficiency,especially later in the season.

Less often,drought injury becomes apparent in the centre of leaves,as in tomatoes,where it appears as angular or irregular leaf spots.Normal autumnal leaf fall of deciduous trees is less a protection against cold than against drought.Leaves of evergreens have anatomical devices to restrict and control transpiration but even in conifers dry spring winds,coming before the roots have begun to function adequately,too often give rise to the well-known " red fir trees."

The terminal buds generally survive on the branches of such trees and later produce the normal shoots with young green needles of the current year,behind which is a bare zone where the needles of one or two previous years have been scorched,dried and fallen off.Premature shedding of flower buds or flowers,especially common in begonias,epi-phyllums and camellias,is often a sign of drought.

In the lily and amaryllis families the flowers react to drought in a peculiar way,by adhesion of the tips of the petals.Fruits also have their own characteristic ways of reacting to a shortage of water.Tomatoes develop blossom-end rot,pears stone cells in the flesh,and apples bitter pit.During severe drought potatoes develop tubers as soft as indiarubber,either at the heel end only or throughout,with a tendency to a sprout in early autumn soon after lifting.

Indirect effects of drought are the encouragement it gives to some insect pests,especially aphides,and to a few parasitic fungi,notably the powdery mildews.Absorption of certain nutrients from the soil is also hindered,especially of minor elements like boron.Such deficiency diseases as heart rot of sugarbeet and canker or black spot of garden beet are thus often associated with dry weather.The boron-deficiency diseases of apples are similarly encouraged and were hence known as " drought spot" before their primary cause was understood.

Because of the severe losses so often caused by drought various methods of artificial water supply have been devised,but there are other less direct ways of mitigating the effects of dry weather.These include destruction of weeds,keeping the surface soil loose by hoeing,planting shade crops or supplying artificial shade,mulching the soil,etc.When plants have been raised under glass it is important to harden them off gradually if they are later to be exposed to a dry atmosphere.

Waterlogged

Roots must be able to breathe,and in waterlogged soil the fine absorbing roots of most plants are choked and die so that absorption of nutrients and even of water is impeded.Water plants have a system of internal air channels which supply the roots with oxygen from above the water level.

Potato tubers are very sensitive to waterlogging and may be killed after a day's complete immersion in water.Occasionally tubers planted in wet soil do not rot but become glassy,very hard,and fail to sprout.Detrimental chemical changes may take place in saturated soil and the population of soil microorganisms may also be adversely affected.Changes of this kind can only be slowly corrected after thorough drainage.Besides producing the suffocated,brown,sour-smelling roots found in waterlogged soil,excessive water supply,whether in a plantation or a flower pot,is often manifested by yellowing of the leaves.

This indicates,amongst other things,that the nitrogen in the soil is not being absorbed by the suffocated roots and partly that it has been washed or leached out of the soil and lost.In house plants this " drowning " sometimes induces a bright yellow colour in the foliage (aurigo),especially in dracaenas and the like.An instance of high death-rate among fruit trees and bushes in Kent is described on Often,especially in greenhouses,the soil contains only a moderate amount of water but,owing to inadequate ventilation,the air becomes overcharged with moisture.

If this condition persists the check to normal transpiration leads to a number of peculiar outgrowths on the above-ground parts of the plants.Oedematous eruptions composed of greatly enlarged cells full of sap appear on vines and tomatoes.In potatoes grown on benches under glass this swelling and eruption of the leaf cells may occur to such an extent that the whole leaf tissue turns mealy and falls away in patches leaving large holes,while the whole leaf becomes curled and deformed.

Such a condition appears very alarming but with sunshine and proper ventilation fresh normal leaves are quickly produced and the plants recover.On leaves of *Pelargonium zonale* wart-like intumescences may be produced by enlargement of the cells which ultimately collapse and die leaving a scab of corky tissue.

The inner surface of pea and bean pods is often covered by a layer of such inflated cells,looking like a coating of white mould.Cacti may react to excessive humidity by becoming in places clear and glassy.In damp situations the breathing pores or lenticels in the bark of many trees become enlarged into corky warts.The same kind of lenticel proliferation may happen in potato tubers produced in wet,though not completely saturated,soil and is often mistaken for a parasitic infection.

In some trees the loosening of the texture of the bark may break out in blisters or involve large patches of the trunk or branches which become covered with a loose corky powder the colour of tan,called by the Germans " tan disease" *(Lohkrankheit).*The condition is probably harmless and may disappear when the trees are transplanted to drier situations but it is disfiguring and hence reduces the market value of nursery stock.

It is advisable to be very careful in winter-spraying trees which show these symptoms.When several big branches are removed together from an old tree the pressure of the rising sap within the trunk may become so high that cracks appear in the bark.It is therefore best to remove undesirable branches a few at a time.An excess of water in the soil favours attack by many parasites,especially where a motile swarm-spore stage occurs in their life history.

Hence powdery scab in potatoes occurs mainly in ill-drained soils and is only prevalent in the wetter parts of the country and on heavy soils; club root of crucifers is also encouraged by a waterlogged soil or a heavy spring rainfall.Some bacterial diseases such as potato black-leg are also favoured by a high soil-moisture content; black-leg is usually only prevalent in ill-drained parts of a field.High atmospheric humidity,followed by condensation of moisture on the roof and dripping on to the plants,frequently leads to an outbreak of grey mould *(Botrytis)* diseases in glasshouses and conservatories. Even out of doors the still damp air of warm muggy days favours an outbreak of many diseases,especially if it is accompanied by dew at night.Well-known diseases encouraged in this way are chocolate spot of beans and potato blight.Sudden changes in water supply from drought to great humidity damage many garden products.

Splitting in plums and other tree fruit is all too common after heavy rain so also are cracked gooseberries,fallen hazel nuts,split tomatoes,telescoped apples,cracked heads of cabbage and carrots and parsnips split right to the core.The potato tuber seems particularly unfitted to cope with a sudden increase in the water supply.In mild cases the skin splits all over in a fine network of cracks,but often the splitting goes much deeper and opens the way to various tuber rots.Splitting generally takes place before lifting but it may also occur during or after lifting.

Sometimes a clear transparent patch of tissue develops at the centre of the tuber which soon dries and cracks to form a lens-shaped hollow,lined with dead brown tissue (hollow heart).This condition is most prevalent in big tubers and is often associated with heavy nitrogenous manuring and wide spacing of the plants.Close planting,balanced manuring and thorough and early cultivation of the potato bed should prevent damage of this kind.

Boron deficiency or infection with some viruses may also increase the tendency to splitting.When heavy rain follows drought towards the end of the growing season potatoes are very liable to second growth.The already-formed tubers develop swollen ends or fantastically shaped outgrowths and the stolons produce an abundance of small new tubers,often several in a row; this last condition is also,however,often associated with the " wilding " condition irrespective of changes in water supply.

The condition known as Jelly End Rot of potato tubers fc apparently due to the roots continuing to function and supplying water to the tubers after the haulm has died or been killed.Such tubers grow out at the rose end and withdraw starch from the tissue at the heel end which becomes glossy and ultimately decays.Under glass a sudden change from drought to high humidity leads to oedema,bitterness in cucumbers,breaking of chrysanthemum stalks just below the head and splitting of the calyx in carnations.

BIO-INSECTICIDES AND PESTICIDES

Today the rapid increase in population and demand of food materials has initiated the large use of insecticides and pesticides.These toxic chemical insecticides and pesticides are resulting in harmful effects and biomagnifications due to which our environment is continuously polluting and fertile lands are acquiring infertility.No doubt they are providing hopeful results in eradication of insects,pests and diseases but they are also killing useful organism present in soil due to which the fertility of the soil is rapidly declining.The conventional farming practices which uses chemical methods to kill both useful and harmful life forms indiscriminately,resulting in the malfunctioning of food chain and food web.Bio-control is the best method to cope with the losses done by the chemicals. In these method insects,pests and pathogens are removed using biological methods without harming the environment and other organism.This is based on natural predation rather than introduced chemicals.The use of bio-insecticides and pesticides also comes under this category.Today due to awareness about the harmful effects of the chemical insecticides and pesticides,most of the farmers are diverting towards the organic farming.In our local area many such plants,waste matter etc.are available from which these bio-insecticides and pesticides can be prepared by using natural means only.Conventional pesticides are generally synthetic materials that directly kill or inactivate the pest.Being single chemical entity,chemical pesticides have resulted in increased resistance in pests.Bio pesticides also known as Biological Pesticides are pesticides derived from natural materials as animals,plants,bacteria,and certain minerals.Bio pesticides are less toxic and also reduce the pollution problems caused by conventional pesticides.

The use of bio-insecticides and bio-pesticides also fall under this category only.Organic agriculture is a unique production management system which promotes and enhances agro-ecosystem health,including biodiversity,biological cycles and soil biological activity,and this is accomplished by using on-farm agronomic,biological and mechanical methods in exclusion of all synthetic off-farm inputs.Organic farming generally produces somewhat lower yields but sustains better yields during drought years,allowing it to reap higher yields in some cases. Studies thus far have shown that organic farming requires less water,uses few and always natural pesticides,prevents soil erosion,leaches dramatically fewer nitrates,and has been shown to have improved nutrient qualities including as much as double the flavonoids,an important antioxidant.This project is mainly based on the preparation,usage,effects and advantages of bio-insecticides and bio-pesticides.

Different types of bio-insecticides and bio-pesticide:

- Cow Urine
- Fermented curd water
- Dashparni Extract
- Neem -cow urine extract
- Mixed leaves extract
- Chili –garlic extract
- Broad spectrum formulation 1
- Broad spectrum formulation 2

COW URINE EXTRACT

In India,as farming goes hand in hand with cattle rearing,therefore the cow urine is available easily in the rural areas.It has acquired an important place in the hindu religion since ancient times.Any religious ceremony and rituals can’t be performed without cow urine which is commonly known as gomutra.We will here use the cow urine for the control of pests and as a growth promoter for the growing crops.

Preparation and Application

Collect 5 liter of pure and concentrated cow urine and dilute it with 40 liters of water and spray it using the sprayer in one hectare of field at the evening time.

Effects and Advantages

- Due to high content of urea in it which is toxic to most of the organisms,the pests and insects etc.will not attack the leaves and buds of the crop plants.
- Due to pungent and bad smell of the extract most of the pests and insects which are attracted due to nectar and fragrance get repelled,preventing the plant.

Fermented Curd Water

In many states of rural India milk production is the main occupation associated with the farming also known as dairy management.The dairy products can also prove a major alternative in controlling the insects and pests in farmlands.One of the dairy product known as butter milk or Chaach can be used for this purpose.

It is made by the centrifugation of curd in which byproduct ghee (fats) is also obtained. There is much greater demand of butter milk in the summer season and it gets distasteful if it is kept for more than two days.This distasteful butter milk can be used to make the extract.

Preparation and Application

Take the distasteful butter milk and add equal amount of water in it.Keep it for two days in a semi-shaded place. Now take it and add 40 liter of water in 10 liter of the extract to form 50 liter of the solution.Spray this solution in 1 hectare of field in such a way that all plants get bath in the fogging spray in the early morning time.

Effects and advantages:

- The fermented butter milk contains billions of bacteria that suppress the growth of other bacteria,fungi and protozoa.The inhibitors released by these bacteria works against the unwanted pathogens.
- Moreover,fermented butter milk also contains the useful nutrients for the growth and maturation of the crop plant.

DASHPARNI EXTRACT

Neem leaves 5 kg,vitex negundo leaves 2 kg,Aristolochia leaves 2 kg,papaya (carica papaya) 2 kg,Tinospora cordifolia leaves 2 kg,Annona squamosa (Custard apple) leaves 2 kg,Pongamia pinnata (Karanja) leaves 2 kg,Ricinus communis (Castor) leaves 2 kg,Nerium indicum 2 kg,Calotropis procera leaves 2 kg,Green chili paste 2 kg,Garlic paste 250 gm,Cow dung 3 kg and Cow Urine 5 lit. Crush all of the above material and mix it in 200 liters of water to ferment for one month.Shake this mixture regularly three times a day.Filter it after one month.This extract can be stored up to 6 months.

Application: Take this extract in fogging machine and apply it on the plants.The above prepared extract is sufficient for one acre crop.

Effects and Advantages

- The extract prepared is very useful for the control of wide variety of pests such as Thrips,Leaf folder,Leafhopper etc.
- It contains neem which has ovi-position deterrent effect,cow dung and cow urine which has bio-fertilizer role also helps in enhancing useful microbial activity in soil.
- It can be prepared easily and can be stored for a large time interval for further use.

Neem-cow Urine Extract

Materials Required: 5 kg of neem leaves,5 liter of cow urine,2 kg of cow dung,100 liter of water

Method of Preparation

Crush the above substances and ferment for 24 hours with intermittent stirring,filter and squeeze the extract and dilute to 100 liters of water.Use this extract to fill in the spray machine and spray it over one acre of the crop.

Effects and Advantages

- Neem is a broad spectrum pesticide and even compatible with most chemical pesticides.Neem works by intervening at several stages of the life of an insect.It may not kill the pests instantaneously but incapacitate it in several other ways.
- Neem acts in various ways like as anti-feedant,as repellent,as growth inhibitor and also has ovi-position deterrent effect.
- Cow urine has high content of urea in it which is toxic to most of the organisms,the pests and insects etc.

Mixed Leaves Extract

3 kg of neem leaves,10 liters of cow-urine,custard apple leaves 2 kg, 2 kg papaya leaves,2 kg pomegranate leaves,2 kg guava leaves.

Crush all of the above recommended material and add 5 liters of water in it.Boil the above mixture 5 times after some intervals of time till the mixture becomes half of the initial.Keep it for 24 hours,filter and squeeze the extract.This can be stored in bottles for 6 months.

Dilute the above prepared extract 2-2.5 liters in 50 liters of water.Fill the diluted solution in fogging machine and spray it in one acre area.

Effects and Advantages

- Guava leaves has achieved a special place in ayurveda due to its medical use for curation of some diseases.
- Pomegranate leaves have resistive nature against many insects and pests due to presence of some special compounds in it.

Chili-garlic Extract

Ipomea leaves 1 kg,500 gm hot chili,500 gm garlic,5 kg neem leaves,10 liter cow urine Crush the above recommended material and boil the suspension till it becomes half of the initial.Filter and squeeze the above extract and store it in glass or plastic bottles. Take 2-3 liters of the extract and dilute it with 50 liters of water.Now mix it thoroughly and use it as a foliar spray for one acre of crop.

Effects and advantages:

- Garlic contains sulphur which is an antibacterial.
- Chili has the property to avoid fungal and bacterial infection due to its preservative property.

Broad Spectrum Formulation 1

3 kg fresh crushed neem leaves,1 kg neem seed kernel powder,10 lit of cow urine,500 gm green chilies and 250gm of garlic

Take a copper container and mix 3 kg neem leaves,neem seed kernel powder with 10 liters of cow urine.Seal the container and allow the suspension to be ferment for 10 days.After 10 days boil the suspension,till the volume becomes half of the initial.

Now ground 500 gm of green chilies in 1 liter of water and keep overnight.In another container crush 250 gm of garlic,add 1 liter of water and keep it overnight.

Next day mix the above boiled extract,chili extract and garlic extract.Mix it vigorously and filter it.Store the concentrate in glass or plastic containers.

Dilute 250 ml of concentrate with 10 liters of water and mix it thoroughly.Now use it as a foliar spray on the crop.

Effects and Advantages

- Due to the formation of complex compounds of Cu in this formulation which are poisonous for many bacteria,fungi,protozoa etc.,it is very useful for microbial control.
- The compounds produced by the fermentation of the above extract are poisonous to many micro-organisms and insects.

Broad Spectrum Formulation 2

5 kg neem seed kernel powder,1kg Karanj seed powder,5 kg chopped leaves of besharam (Ipomeas sp.),5kg chopped neem leaves,10-12 lit of cow urine

Suspend 5 kg neem seed kernel powder,1kg Karanj seed powder,5 kg chopped leaves of besharam (Ipomeas sp.) and 5kg chopped neem leaves in a 20 liter drum.Add 10-12 lit of cow urine and fill the drum with water to make 150 lit.Seal the drum and allow it to ferment for 8-10 days.After 8 days mix the contents and distil in a distiller. Distillate obtained from 150 litre liquid will be sufficient for one acre.Dilute in appropriate proportion and use as foliar spray.

Effects and Advantages

- Neem effectively controls common pests like Thrips,Whitefly,Leaf folder,Bollworms,Aphids,Jassids,Pod borer,Fruit borer,Stem borer,Leafhopper,Caterpillars,Diamond back moth.
- Karanj contains poisonous chemicals in it and can be used to check microbial growth.

BIOLOGICAL CONTROL AGENT FOR NEMATODES

Too often,biological control agents have failed because they have been used before a basic knowledge of their ecology and biology has been established.The importance of such knowledge can be seen from work at Rothamsted aimed at the development of *V.chlamydosporium* as a biological control agent for root-knot nematodes.Stirling describes the design and methods used in the conduct of biological control experiments,which are more complex than those required to assess the efficacy of a nematicide.He identified five key aspects in setting up an experiment to evaluate a biological control agent and these are presented below in a slightly modified form.

- The test organism and any organic amendment should be applied at practical application rates; 0.1 percent w/w soil is equivalent to 2.5 tonnes/ha and should represent a maximum dose.Tests should always be performed in a non-sterilized soil with a natural residual soil microflora.
- Appropriate treatments as well as an untreated control should be included if the organism is added with a substrate.These treatments should include the substrate alone,the organism alone,and the autoclaved colonized substrate.Too often,untreated controls are compared only with large applications of the organism and substrate and this does not allow separation of the effects of the agent from the effects of the substrate.In several tests reported in the literature,application of the substrate alone has decreased nematode populations to the same extent as the substrate colonized by the agent,and there is no clear evidence of biological control.

- Population densities of the agent under test should be monitored to ensure that it has survived in soil throughout the period that activity against the nematode target is required.Such monitoring may require the development of selective media,which can be a difficult and time-consuming task.
- Nematode mortality caused by the organism under test should be measured to assess whether differences between nematode population densities in treated and untreated soil relate to the levels of kill caused by the agent.Infection levels are relatively straightforward to estimate for most parasites,but repeated sampling is required to determine total kills.The effects of agents which produce toxins or have indirect effects on nematodes through competition,the modification of root exudates or the colonization of feeding cells,can only be measured by assessing their impact on nematode development.
- The impact of the soil environment,host plant and nematode should be tested as these are likely to affect the efficacy of the biological control agent,and could account for the lack of activity of potential agents in specific test conditions.

In a review of the literature less than 15 percent of experiments purporting to demonstrate biological control caused by *Paecilomyces lilacinus* satisfied the above criteria.Although this is an unsatisfactory situation that must be remedied,the difficulties in conducting carefully controlled and monitored experiments should not be underestimated.

Isolates of *V.chlamydosporium* differ markedly in their growth and sporulation *in vitro*,and in their virulence,saprophytic competitiveness and rhizosphere competence.Such differences between isolates of the same species of micro-organism are common and there is a need for simple laboratory-based screening methods to select the most promising isolates for further testing.*Verticillium chlamydosporium* isolates were collected from infested nematode females and eggs in suppressive soils around the world.Hence,they were isolated from the niche in which they would be required to be active as a biological control agent and from soils most likely to yield active isolates.Once isolated in pure culture using standard techniques,the *in vitro* growth requirements were determined.

Virulent isolates were selected by counting the number of nematode eggs parasitized after exposure to the fungus on agar in a standard test.Tests for proliferation in the rhizosphere and growth in soil were also used to select promising isolates.Although simple laboratory-based screens help eliminate many isolates that show insufficient activity to justify further testing,selected isolates will not necessarily be active in the field.As it may take 10 to 16 weeks to investigate adequately the performance of different isolates against cyst and root-knot nematodes in pot tests,relatively few can be screened.The method of mass culturing of *V.chlamydosporium* for experiments can have a marked effect on the subsequent survival and proliferation of the fungus in soil.Inoculum produced in shaken liquid cultures consists mostly of hyphae and conidia,which require an energy source to ensure proliferation in soil,whereas on solid media large numbers of chlamydospores are produced and these can be added to soil in aqueous suspension and rapidly establish the fungus.

The development of a semi-selective medium has enabled detailed studies to be made on changes in relative abundance of the fungus in soil and on roots.Some isolates of *V.chlamydosporium* may be extremely abundant in soil but unless they are capable of colonizing the rhizosphere they do not parasitize the eggs of root-knot nematodes.

Growth in the rhizosphere differs markedly between plant species,*e.g.*,tomato,cabbage and maize roots support much growth,whereas sorghum,pepper and cotton are poor hosts.Colonization by the fungus is confined to the rhizosphere and rhizoplane and there is no spread into root tissue; no lesions have been observed on roots grown in soil treated with *V.chlamydosporium* and there have been no detrimental effects on the growth of a range of crop species.

A tenfold reduction (from 10^4 to 10^3 chlamydospores/g soil) in the amount of fungus applied to soil had no effect on the extent of colonization in the rhizosphere; the ability to proliferate on the root surface where the fungus is required to control nematodes is an important characteristic which may allow significant reductions in the amount of inoculum applied to soil.

The efficacy of *V.chlamydosporium* as a biological control agent for root-knot nematodes is affected by three key factors: the amount of fungus in the rhizosphere; the rate of development of eggs in the egg masses; and the size of the galls in which the female nematodes develop.In large galls female root-kn2ot nematodes may produce egg masses which remain within the gall and are not exposed to parasitism by *V.chlamydosporium,* which is confined to the rhizosphere.Hence,*V.chlamydosporium* is less effective in controlling root-knot nematodes in heavily infested soils and on highly susceptible crops because large galls are formed on the roots and many eggs escape parasitism.

Verticillium chlamydosporium is unlikely to be useful in these situations where a grower would normally apply a nematicide.Also,at temperatures above 25°C eggs may complete their embryonic development and hatch before the fungus has completely colonized the egg mass; at 30°C about 30 percent of eggs of three root-knot species hatched and the second-stage juveniles escaped from the egg mass before the eggs were killed.These studies in pot tests,if supported by field experiments,help to define the conditions in which *V.chlamydosporium* might be used successfully for control of root-knot nematodes.It is only from such detailed studies that the limitations and requirements of the fungus can be assessed.

The Extent of Colonization of the Rhizosphere by *Verticillium chlamydosporium* on Several Plants Grown on Soil Treated with 5 000 Chlamydospores/g Soil and the Control of *Meloidogyne incognita*

Verticillium chlamydosporium is not a replacement for nematicides but,despite its limitations,it may be a useful management tool.In most tests biological control agents have been applied to protect susceptible crops.However,application of the fungus to a relatively poor host for the nematode on which small galls are produced so that most egg masses are exposed in the rhizosphere might provide more effective control; soil population densities of the nematode would be reduced to non-damaging levels before a susceptible crop was planted.

Bridge recommends the following rotation for the management of root-knot nematodes: susceptible host - poor host - poor host - resistant or non-host -susceptible crop.Application of the fungus before the first or second poor host may permit the shortening of rotations,or the replacement of the resistant or non-host crop,without increasing nematode damage to the susceptible crops.Currently,at Rothamsted Experimental Station,United Kingdom,poor hosts for root-knot nematodes are being screened for their ability to support *V.chlamydosporium* in their rhizospheres.It is hoped that by combining the host plant and the biological control agent,more effective and consistent control can be achieved.

The Next 20 Years

The past 20 years have seen a significant increase in the number of scientists involved in research on the biological control of nematodes.Surveys and empirical tests are being replaced by quantitative experimentation and basic research on the modes of action,host specificity and epidemiology of selected organisms.Such basic information is essential for a realistic appraisal of the impact of molecular biology on the improvement of microbial agents and monitoring the spread and survival of released organisms,and for the development of rational strategies for control.

Current experience suggests that biological control agents will not replace the use of nematicides but,integrated with other control measures including chemicals,they could play an important role in the development of integrated control strategies in both developed and developing agriculture.The urgent need to reduce the dependence on nematicides should provide the necessary impetus for the considerable amount of research and development still required to ensure the successful use of such agents.

Pathology

Bailey and Gordon reported that *R.culicivorax* depleted host metabolites,reduced fat body and other host storage tissues while accumulating stored materials in their trophosomes,and thus inhibited the development of imaginal disks in the host mosquito.Hemolymph proteins were depleted sixfold in mosquitoes.

Mermithid parasites of larval diptera probably resemble *M.nigrescens* in obtaining dietary amino acids by stimulating the catabolism of proteins within the host fat body.In addition to the reduction of most protein fractions,significant decreases also occurred in hemolymph glucose in parasitized blackflies; however,blood trehalose concentrations were not affected.Glycogen reserves were shown to be similarly reduced in the fat body of mermithid- parasitized simuliids.Gordon *ET al.*found that parasitism did not significantly affect either the overall concentration of lipids or relative proportions of the lipid fraction in the hemolymph of mosquitoes.Mermithid parasitism of mosquitoes and simuliids caused almost complete degeneration of host fat body tissue.All storage metabolites,including glycogen,within the fat body may directly or indirectly be utilized by the developing nematode.

Responses are somewhat different in larger hosts with longer periods of parasitism.Protein in the fat body of grasshoppers parasitized by M.*nigrescens* was depleted after 2 weeks; corresponding reductions in proteins were not recorded from the hemolymph until I week later.The third week of infection represented a period of maintenance and reconstruction of fat body soluble proteins,but corresponding effects were not found for hemolymph proteins until the fourth week of infection.

Gordon and Webster found that the overall level of amino acids in the hemolymph of adult female grasshoppers was not affected by mermithid parasitism,but the content of amino acids within the fat body was significantly reduced.Parasitism by *M.nigrescens* resulted in glucose levels remaining low for the first 4 weeks of parasitism relative to trehalose,but no significant difference for glucose levels when compared to unparasitized controls.Trehalose levels dropped and remained low for much of the parasitic period.These data suggested that the nematode modifies its host's metabolism to favour production of small molecules such as glucose and amino acids from carbohydrate and protein reserves and that the host cell's ability to sequester nutrients is impaired during parasitism.

Craig and Webster found that ecdysone levels in locusts were unaffected by *M.nigrescens* and attributed the inhibition of molting in parasitized insects to depletion by the nematode of precursors required by the host for protein and cuticle synthesis.However,ovaries of parasitized locusts were unable to sequester vitellogenic proteins available within the hemolymph,suggesting endocrine dysfunction in the host.

Major modifications of the host's metabolism by mermithid parasitism are manifested in host tissue degeneration and retarded development including resorption or suppression of oocyte development.As a result,hosts are usually prevented from maturing to the adult stage or when they do they are generally unable to-,reproduce.

Hosts are occasionally able to prevent parasite development by either cellular or humoral responses.The most common cellular response is encapsulation.Poinar *et al.*reported that *Culex territans re*sponded to the development of parasitic juveniles of *R.culicivorax by* covering the 2- to 3-day-old parasites with blood cells.These nematodes failed to complete development.A more frequent condition,melanotic encapsulation,was reported in *Diabrotica* beetle larvae attacked by F*.leipsandra.*Host hemocytes lysed on the cuticle of the nematode soon after it entered the host,and within 6-8 hours an inner layer of melanin had formed around the nematode.

Similar responses were observed for larvae of *Psorophoraferox,Aedes triseriatus,and Anopheles quadrimaculatus* to the *R.culicivorax.How*ever,some 70 species of mosquitoes have not elicited this type of defence response to *R.culicivorax*.Recently,Gaugler *et al.*reported an encapsulation response when *Empidomermis* sp.entered larvae of *Aedes stimulans* if the nematodes were unable to migrate to the host's head and enter central nervous tissue.

After pupation,the nematodes dropped back into the hemocoel where they no longer elicited a host defence reaction.This response may be common in mermithids that enter the larval stage but fail to mature until the host reaches the adult stage.Also,Harlos *et al.*noted that *Culicimermis* sp.entered the nerve tissue of *Aedes vexans* larvae and later developed in adult hosts; Petersen (unpublished data) observed a similar response in *Ae.sollicitans* when parasitized by *P.culicis.*

Humoral responses,non-cellular components of the insect's hemolymph which have an adverse effect on nematode development,may account for the host resistance observed in *Aedes intrudens* and *Aedes provocans* to an *Empidomermis* sp.when no discernible host response was observed,which was unlike the encapsulation response this nematode elicited in *Ae.stimulans*.Another type of humoral response may be that observed when blackflies were subjected to the mosquito mermithid *R.culicivorax.*The mermithid initiated development in early-instar; Iackflies,and direct host reaction to this unnatural parasite was not observed; eventually,however,both nematode and host died.

Physical and behavioural responses also influence parasitism by mermithids.Host age,especially in aquatic insect hosts,has been shown to have a considerable effect on host susceptibility.Refractiveness because of age appears to be caused by the thicker cuticle in older larvae.In the laboratory,as a result of mass-rearing practices for *R.culicivorax,* where exposed but uninfected hosts were retained to supply the next generation,the host mosquitoes were found to have become measurably less susceptible to the nematode after some 100 generations.No evidence of humoral defence mechanisms was found and the mode of defence was not determined.In a similar buildup of refractiveness in *An.quadrimaculatus to S.Petersen,* it was reported that less susceptible hosts were more active and aggressive in attempting to remove attacking preparasites.

Natural Population Regulation

Mermithids generally have localized,discontiguous distribution.This is especially true for mermithids that develop in larval hosts.*Romanomermis culicivorax* was found in only seven pools out of several thousand examined in southwestern Louisiana.This is understandable because the hosts are killed prior to emergence from the habitat,thus greatly reducing the nematodes ability to disperse.

Generally,observations of mermithid parasitism are based on isolated collections and often give the impression of higher parasite activity than actually exists.In extended studies,levels of parasitism are usually found to vary greatly.In a 27-month survey of five habitats containing populations of *R.culicivorax,* mean parasitism of *Anopheles crucians* for individual habitats ranged from 8 to 42 per cent,with parasitism of individual populations as high as 80 per cent.Similar but lower levels of parasitism were observed for several other mosquito species. Similarly,Phelps and De Foliart observed parasitism of blackflies,*Simulium vittatum,* in four Wisconsin streams over a 1-year period.Parasitism typically ranged from 10 to 90 per cent and the authors suggested that localized parasite populations probably attained levels of parasitism sufficient to virtually eliminate some host populations.Similar population reductions of blackflies were reported by Welch and Rubtsov.

Parasitism of adult aquatic insects usually occurs over a wider geographical area and often reaches very high levels but fluctuates greatly over time.Steiner observed 80 per cent parasitism of adult *Ae.vexans in* British Columbia in 1920,but only 20 per cent the following year.Trpis *et al.* reported 100 per cent parasitism of the same host in the same locality in 1967.Similar observations were made on parasitism of *Ae.sollicitans* adults by *P.culicis* in southwestern Lousiana.During an entire year only collections made in January failed to produce infected hosts; mean monthly parasitism ranged from 0 to 48 per cent with individual populations reaching 96 per cent.During a period of high parasitism,a survey was made along 80 km of coastline and revealed parasitism ranging from 48 to 94 per cent for seven populations.

Similar isolated high levels of parasitism are frequently observed in terrestrial insects.Christie reported a heavy outbreak of grasshoppers in Wisconsin in 1923-1925 with a correspondingly heavy infestation *of M.nigrescens.* The author concluded that the parasites were an important factor in terminating the grasshopper problem.Similarly,Mongkolkiti and Hosford reported that a population of the grass*hopper Hesperotettix viridis pratensis* was totally destroyed prior to egg lying by *M.nigrescens.*Other species of grasshoppers in this population were only partially affected or not affected by the mermithid.

Though reports of natural insect control are impressive and suggest the potential of mermithids as biological control agents,these parasites generally do not influence significant control over host populations.In a review of blackfly-mermithid literature,Molloy reported that parasitism in most blackfly populations is moderate,ranging from 3 to 15 per cent,and is perennial with only rare and highly localized epizootics.This pattern probably holds true for most mermithid-host associations.

Safety

As previously mentioned mermithids are generally specific to one or a few species of insects and are rarely found in other insects.As a result,little concern has been shown for potential danger of these parasites to mammals.To date,only limited mammalian safety studies have been made.Ignoffo *et al* reported that when suckling and adult mice and adult rats were subjected to either *per os,*intranasal,intraperitoneal,or dermal challenge of *R.culicivorax,*their body weight gain and histologies were identical to those of untreated animals.Immunodepressed rats also were not susceptible.Similar findings were reported for *R.iyengari in* India.Similarly,studies in the People's Republic of China showed that no pathology was apparent in dermally challenged suckling mice,in orally exposed adult rats,or in three species of fish when exposed to *R.jingdeensis*.

Mermithids have occasionally been reported as accidental parasites of man.Poinar reviewed all such cases in the literature and concluded that in most cases the data were insufficient to ascertain if human parasitism actually occurred.He further concluded that the question of accidental human infection by mermithids cannot be definitely answered at this time and should only be accepted as fact when proven experimentally or when parasites are found developing *in situ* in the human organism.Further,the Environmental Protection Agency has exempted mermithids from safety regulations.

Mass Propagation

Mass production systems at economical costs are essential if a given biological control agent is to be effectively used.Mermithids take in their nutritional,needs directly through the cuticle,making them especially difficult to propagate by *in vitro* methods.Therefore,pest control attempts have been limited to those mermithid-host systems that are conducive to *in vivo* rearing.

In Vivo Propagation

The *in vivo* rearing of mermithids of terrestrial insects is especially difficult because of the protracted life cycles of many of the hosts.Though numerous mermithid species have been encountered,laboratory culture has been limited to *M.nigrescens and F.leipsandra.* Laboratory infections are relatively easy with *M.nigrescens.* Gordon and Webster collected adult females of *M.nigrescens* from the field,allowed the females to oviposit,concentrated the eggs on moist filter paper,and stored the eggs at 5°C until needed.To infect hosts,eggs were transferred onto small pieces of grass previously coated with an adhesive material.The grass was readily eaten by host grasshoppers.This method permitted extensive research on the physiological effects of mermithid parasitism on a host system.The long developmental period of the nematodes during both the parasitic and free- living stages (I year) has prevented meaningful use of this nematode in field trials.

Creighton and Fassuliotis were the first to give a detailed report of the laboratory culture of a mermithid of terrestrial insects.Preparasitic juveniles of *F.leipsandra* and the larvae of the cucumber beetle,*Diabrotica blateata,* were placed in holes in soil in clay pots.The pots were sealed and held for parasite emergence.Ratios of 1000 nematodes per 100 host larvae were sufficient for maximum recovery of postparasitic nematodes.The method has the potential of producing between 80,000 and 240,000 nematode eggs per experimental unit.From this basic procedure,Creighton and Fassuliotis developed an *in vivo* mass rearing system for *F.leipsandra.* As eggs were laid,they were concentrated,sterilized by immersion for 10 minutes in 0.28 per cent sodium hypochlorite,and rinsed.

The eggs were transferred to a Baermann funnel and allowed to mature and hatch.First instar larvae of the banded cucumber beetle were exposed to the preparasites in clay pots as previously described.Twenty-four pots were placed in a plastic box and held 4-5 days at 24-27°C.The contents of the clay pots were then transferred to porous baskets.The baskets were placed at one end of a larger container with sand at the other end.As the host larvae matured,they migrated to the sand and pupated.After about 10 days the nematodes began to emerge from their hosts.The postparasites were then transferred to a container of aerated water until the nematodes molted to the adult stage and began to deposit eggs (7-10 days).Each larger container (37 x 27 x 10 cm) produced an average yield of 1.9 X 106 nematode eggs.The system can produce about 5 x 106 eggs week- I at a cost of $0.40 (US$) per I 000 eggs including supplies and labour for producing the host insects.

Although mermithids are relatively common in aquatic insects,especially in the Simuliidae,Chironomidae,Ceratopogonidae,and Culicidae,many of these host species cannot be successfully maintained in laboratory colonies or cannot be maintained in large numbers.Also many mermithid species undergo egg diapause or have an asynchronous hatch which prevents their mass production.As a result,few aquatic mermithid species have been studied extensively.The midge parasite,*Hydromermis conopophaga,* has been readily maintained in colony because the host,*Tanytarsus* midges,can be easily maintained.However,*H.conopophaga* has never been mass produced or released in the field.Similarly,the mermithid *Limnomermis rosea,* a parasite of chironomid larvae,has been reared through several generations but has not been released in the field.

Muspratt was the first person to develop,maintain,and report a laboratory culture of a mermithid nematode.He collected *0.muspratti (Agamomermis)* from naturally infected host mosquito larvae and placed the nematodes in containers filled with moist sandy soil.The soil was allowed to dry out over a period of several months.After 11-12 months some of the soil was placed in a container of water and first instar mosquitoes added.Seventy to 80 per cent of the host larvae contained nematodes.The original work by Muspratt though very basic provided the direction for future work with mosquito mermithids.Subsequent studies with 0.*muspratti* have refined the rearing technique and defined biological limitations.A mass production system for *0.muspratti has* failed to develop,even though a suitable laboratory host system is available,because the eggs of this species failed to hatch at the same time although they appeared to mature.Because of the non-synchronous hatch,cultures were observed to produce infective stage nematodes after periodic floodings for over 5 years (Petersen,1981).

The mermithid,*S.peterseni,* a parasite of *Anopheles* mosquitoes,has been maintained in colony for over 10 years.Though this nematode appears to be an infective parasite,and is easily maintained in laboratory colony,economical mass production systems have been difficult to achieve because the only hosts are *Anopheles mos*quitoes,which are less suitable for mass rearing systems than are certain *Aedes and Culex* species.

The most efficient and effective *in vivo* rearing system for mermithids is that developed for *R.culicivorax.*The primary techniques were developed in 1971.This initial system called for the exposure of 20,000 first-instar *Cx.quinquefasciatus* larvae in 136 x 52 x 5 cm galvanized 'Lrays to preparasites at a parasite-host ratio of 12: 1.The hosts were then fed a regimented diet for 7 days.After this period the culture trays were drained and the mosquitoes concentrated.Pupae of uninfected mosquitoes were separated using a chilled water procedure,and the infected mosquitoes were placed in 36 x 25 x 10 cm trays and held for nematode emergence.

Postparasitic nematodes were then washed and 10-15 g placed in paraffin-coated aluminum pans (22 x 33 x 5 cm) which contained clean,coarse,sterile sand and water.After 3 weeks the free water was removed and the cultures stored an additional 4-15 weeks before use.When preparasitic nematodes were needed the cultures were flooded with chlorine-free water to stimulate nematode hatch.Nematode cultures averaged 1.91 million preparasites (45-60 per cent of total yield) if they were flooded for the first time when they were

11- 19 weeks old.Total yields were highest (5.32 million preparasites) when cultures were flooded for the first time when 8-10 weeks old and then at 3-4 week intervals thereafter.

It has since been determined that yields can be increased by having a density of 24 postparasites CM-2 in the culture trays.Yields of preparasites were tripled by simply setting up three cultures (22 x 33 x 5 cm),each containing 5 g of nematodes,instead of the usual method of one culture containing 15 g of postparasites.The longevity of cultures could be extended for up to 9 weeks if the cultures were allowed to mature at ambient temperature and then held at 5-IOC.Also,immature cultures survived best at 15-20'C.Further,Sterling and Platzer reported that by adjusting the pH of water to 4.5 in R.*culicivorax* cultures,the nematophagous fungus *C.anguillulae* could be effectively controlled.

Chapman and Finney reported that mature eggs of *R.culicivorax* can be stored for 12 weeks at IO°C and 21 weeks if stored during early development at 15°C and later transferred to IO°C.They also reported that eggs of *R.culicivorax* placed in a damp fluoroform product could be shipped with only a 10 per cent loss.The technique has the advantages over the standard shipment of sand cultures of being about eight times lighter,and costing less to ship in addition to reducing shipping losses.

In an actual test of the mass propagation system,sufficient *R.culicivorax* were reared to treat 144,000 m2 of breeding area in El Salvador.The necessary inoculum required the exposure of 1.6 million first-instar *Cx.quinquefasciatus* to 137 million preparasites (I: 14 ratio) each week for 6 weeks.The system produced an average of 13.7 g (about 2000 g- 1) of postparasitic nematodes per rearing tray (20,000 mosquitoes),a total of 6392 g for the 6-week period and 435 cultures.

In Vitro Propagation

The main attribute of successful *in vitro* rearing systems is that they will permit the production of species -not currently available by *in vivo* methods.Also,*in vitro* systems offer the potential for reduction of costs over *in vivo* systems,and better product standardization and formulation.The development of a successful *in vitro* rearing system for mermithids has been slow because of a lack of readily available material.Consequently,most basic *in vitro* research has been done with *R.culicivorax.* Another major problem involves the nature of food ingestion in mermithids.

Unlike most nematodes,mermithids lack a functional gut and must rely on transcuticular uptake of nutrients.Since these nematodes feed only during their parasitic phase while in the hemocoel of the host,nutritional products must be in a form that can be readily absorbed through a selective and delicate cuticle.Thus,they are very sensitive to mechanical damage and to hypo- and hypertonic changes. In initial studies with *R.culicivorax, Sanders et al.* reported growth of 1-4 mm in length after 15 days and 5-7 mm after 25 days and some stichosome and trophosome development when preparasites were placed in Schneider's *Drosophila* medium with 10 per cent fetal calf serum at pH of 6.5-7.0 and 25°C.Growth rates were about one-third that of nematodes grown *in vivo*. More recently,Castillo *et al.* tested more than 50 combinations of vertebrate and invertebrate tissue culture media and microbiological media.Slow growth and limited development of internal structures were obtained with various supplemented Grace's tissue culture media and Schneider's *Drosophila* media.Nematodes attained a stage of development after 3-4 weeks comparable to that attained in the host after 4-5 days.

The most successful attempts to rear *R.culicivorax in vitro* were reported by Finney.She reported that the best results were obtained using Grace's medium containing 10 per cent fetal calf serum at 26°C with osmotic pressure adjusted to 240 mosm and pH 6.4-6.5.Slow growth of the nematodes occurred over a 6-week period during which only juvenile females developed,but storage material in the trophosome was lacking.Finney has since reduced the time to reach parasitic maturity to 3 weeks with the trophosome filled with material.However,she was still unable to produce male nematodes.

METHODS TO CONTROL PEST INSECTS AND MITES

Other methods to control pest insects and mites,such as introduction of attractant and repellant materials,application of growth-regulating chemicals,or mass release of pheromones to cause mating disruption,which have been tested or applied to other crop systems,have not appeared promising or have not been assessed for controlling pests on vegetable crops.

Similarly,the release of sterilized males to reduce development of pest species has not proven economically practical for control of insects on vegetables.

In greenhouse vegetable crops,the high degree of environmental control available to growers enables fully integrated programmes to be devised that give good productivity as well as control of insects,mites and diseases.Biological control,using natural and introduced parasites and predators of pathogens and pests,plays a major role in greenhouse crop protection,but it is more difficult to realize in field vegetable crops because of the lack of control over the environment.

Management of viral diseases depends largely on vector control,crop hygiene,resistant cultivars,and regulatory systems of quarantine,inspection and certification.

Monitoring Certain Pests

Monitoring involves some form of surveillance to detect the presence of a disease or pest before economic damage has occurred and in time to economically apply eradicative or protective measures.Two such methods involve visual inspection and trapping.Visual inspection is the simplest method of monitoring and is used for insect and mite pests and for certain diseases,especially in combination with weather data and epidemiological information concerning sporulation,spore dispersal,infection periods and host plant growth stages.Surveillance is done on a regular basis,beginning soon after seeding or planting out,and includes watching for signs of feeding or irregularities of plant growth.

Trapping is necessary for monitoring certain pests; traps may be purchased for certain pests or some may be constructed at home.The type of trap and its location must be suitable for the target species.Sticky strips or glass slides smeared with petrolatum or other adhesive are used for trapping fungal spores,aphids and thrips; coloured pans with water in the bottom are used for aphids; coloured sticky strips or ribbon-like tapes are available commercially for whiteflies in greenhouses; a variety of lights and light traps are used for moths of many types; and a pheromone in a rubber septum is available for such insects as the European corn borer.Determining the density in soil of nematode pests before planting is important for some crops.

Threshold levels have been established for a number of vegetable crops with regard to the rootlesion and root-knot nematodes.

Cultural Practices

After genetic resistance,the most important strategy for managing plant diseases and pests is best thought of as escape or prevention rather than cure.Such measures are aimed at avoiding the conditions that predispose the crop to infection or infestation; for example,selection of healthy seed of high germinability from a reputable source; treating seed if necessary to eradicate pathogens; checking the health of transplants at their source; practicing adequate crop rotation; sowing and transplanting only when soil conditions (temperature,moisture,nutrients,tilth) are adequate; maintaining adequate spacing and row orientation parallel to prevailing winds to permit good crop ventilation; avoiding overhead irrigation when the crop is in a vulnerable stage (*e.g.* at flowering in bean and pea); avoiding working in fields when the foliage is wet with dew or rain; roguing infected plants from seed crops; picking fruit before it is overripe; avoiding mechanical damage and removing field heat as quickly as possible at harvest; controlling weeds that may harbour pathogens and arthropod pests and contribute to humid microclimates in the crop; and controlling insects that may cause infectible wounds and transmit viruses.Crop hygiene is a major factor in disease escape because it removes many sources of pathogens.Trash piles,contaminated seed trays,pots and stakes,and infected weeds are important sources of wind- and water-borne fungi and bacteria.Personal hygiene and equipment sanitation are also very important.Bacteria and viruses are easily transmitted on hands,clothing,tools and machinery.Tomato mosaic virus,for example,can persist for several months or years on sapdrenched overalls if they are hung dry in a dark closet.In bean fields,bacterial diseases often follow the tracks of workers walking among wet plants.

In nature,populations of insects and mites are regulated primarily by climate-related factors; by predation and parasitism by other arthropods and by other animals,such as birds; by diseases; by preference for food source; by competition with other plant feeders; and by numerous other factors.Because of the significant mortality of insects and mites by predation,parasitism and diseases,the best control practice for many insects and mites is to do nothing that could reduce the impact of these naturally occurring beneficial organisms and other beneficial agents in the environment.

Cultural practices usually influence the impact of pests indirectly.Tillage,crop nutrition,crop rotation,timing of planting and other management practices affect the vigour of the crop plant.A vigourous,healthy plant is more likely to withstand damage from feeding by sap-sucking insects and mites.Attack by a pest sometimes can be circumvented or damage reduced by tillage,by timing of planting and by selection of field conditions.For example,planting cruciferous crops in well-drained soils and delaying planting to reduce exposure of young plants to cold,damp soil conditions will reduce damage by root maggots.The removal of crop residue from the field and other sanitation procedures may reduce infestations of some pests,such as the European corn borer,in the following crop.

INTEGRATED PEST MANAGEMENT STRATEGIES

Integrated Pest Management strategies require detailed studies which can be broken down into three steps:

1. A precise description of pest population dynamics in space and time in order to assess damage thresholds,to determine key points for control (possibly by modelling),and to evaluate control efficiency;
2. A general survey to estimate the variability in the first step between seasons or across a region;
3. The control strategy,including a survey by the grower of population dynamics.

Each of these steps requires particular sampling methods that differ in accuracy: precise measurements for detailed studies,less precise measurements but which can be used on a larger scale for variability evaluation in the second step,and quick and simple methods for final use by the growers.

Pest and disease intensity may be quantified using two different measurements:

1. Estimation of the population size,*e.g.*number of aphids per leaf,number of fungal spores in a cubic meter of air,etc.;

2. Quantification of the injury caused to the host plant,*e.g.* the proportion of leaf tissue infested by larvae,the relative leaf area covered with disease symptoms,etc.The methods should be easy to use,allow rapid estimation,be applicable over a wide range of conditions,and most of all,be accurate and reproducible.This chapter presents some of these methods which may be used for greenhouse crops.

INSECT PESTS

Estimating Insect Numbers in Samples

We will examine different ways of reducing pest assessment time.Methods based on visual abundance indices will be developed in particular,and examples of their application to insect pests will be given.Since many authors have developed methods which may be used with species other than those that they have studied,the references do not always concern greenhouse species.At each step of a study,the spatial distribution of most pests will be very patchy.Sampling plans will thus require numerous data to reach the required level of accuracy.Particular attention should be given to the evaluation of pest densities at each sampling point.This is a bottleneck which will define the "cost" of the sampling.

Methods are often available to reduce the cost of counts,but,except when automatic counting (*i.e.* picture analysis) is possible,these methods lead to a drastic decrease in accuracy.This loss of accuracy is cumulated with the error induced by the sampling itself,to define the final value of the density estimates.Moreover,several methods that are used to accelerate pest counts (*e.g.* field visual observation) underestimate pest densities per area or volume.Such systematic bias,as well as the accuracy of estimates,must be evaluated before using this kind of method.

The most accurate way of obtaining quantitative data on a pest population is to collect the substrate (*e.g.* host plants) and to take these samples to the laboratory where individuals may be isolated and counted under a stereoscopic microscope.As this method is time-consuming,numerous authors have attempted to reduce the time required.The first step is the mechanical extraction of the individuals from the substrate.They can be extracted by washing,by flotation in high-density medium such as saccharose,or by the use of Berlesse-Tullgren funnels.Once the insects are separated from mud,sand or plant fragments,the clean extract can be fractionated into sub-samples.Both steps give results of varying precision,depending on the medium surrounding the insects and the species involved.The time required for counting the insects is reduced two to five times,but is often still too great (*e.g.* up to two hours per leaf,including washing,sub-sampling and counting,for aphids on cucumber).

Collecting insect substrate is not efficient for species such as thrips,which are very mobile.In this case,the extraction has to be made directly in the field,by using sweepnets,direct picking,mouth vacuum devices or vacuum nets like the Dietrick vacuum (D-vac).The numbers are then calculated from the part of the population which can be recovered.This part is often highly variable and the precision of the method is difficult to evaluate.However,successive sampling at the same sites may enable insect density to be estimated with accuracy.The method of Suber and Le Cren,frequently used to evaluate fish densities in rivers,was adapted to benthic insect counts by Lapchin and Ingouf-Le Thiec,and then by Lapchin to estimate larval and adult coccinellid densities in wheat fields.It has recently been used in cucumber greenhouses by Boll to estimate the number of thrips on leaves after several successive strikes.

Estimating Insect Population Densities

Insect densities may be estimated *in situ,* without collecting samples.The characteristics of the species distribution on the host plants should be taken into account and the most representative leaves or stems should be observed.This method eliminates the laboratory stage of density estimation,but does not significantly reduce counting time.Observation time may sometimes be drastically decreased if rough counting is performed.

This method is useful if the strong systematic under-estimation of the numbers it produces is constant or depends on known parameters.In the previous example of *Coccinella septempunctata* L.in wheat fields,Lapchin developed a "quick visual method",*i.e.* the observer walked within the 25 m^2 sub-plot for 2 min and counted all the adult coccinellids he saw.The numbers obtained by this method correlated well with the density estimated with the Seber and Le Cren method,and allowed the development of a sequential sampling plan for adult coccinellids in wheat fields.However,such a quick method is not automatically appropriate,even for closely related species,as Frazer and Raworth did not find that "walking counts" of adult coccinellids in strawberry fields were reliable.Variance of insect numbers is closely related to the mean.This property implies that variations in numbers may be more easily perceived by observers who follow a geometric rather than arithmetic scale of density.In the field,such orders of magnitude can be easily translated into abundance classes.This was probably the reason why the use of categorical data was soon considered to be a good way of drastically reducing sampling costs.

In the case of aphid populations,precise density estimates of *Myzus persicae* (Sulzer) were related to the proportion of infested leaves in different parts of potato plants.This method was used on other host plants and statistically developed by several authors.As far back as 1954,presence-absence methods were improved by use of a set of abundance classes.Such classes can be purely arbitrary and,for instance,the sampling units distributed into poor,medium or heavy infestation classes.Several authors used more precisely defined classes,according to the number of colonies,their size and their localization.Different kinds of abundance class systems may be developed,according to the insect studied and its environment.

Building a System of Abundance Classes

Three main types of class can be considered.Firstly,there are classes whose limits are defined by the number of individuals that are seen during one sample unit of observation.A logarithmic scale of these limits was first used by Leclant and Remaudière to estimate *M.persicae* densities on peach trees.Another scale which is based on the approximate powers of has successive classes such as: no insect seen,1 to 3,4 to 10,11 to 30,etc.

This scale was used by Ferran to evaluate the density of the rose aphid [*Macrosiphum rosae* (L.)] on rose bushes,by Boll and Lapchin for *Macrosiphum euphorbiae* (Thomas) in tomato greenhouses,and by Lapchin to estimate mummified *Aphis gossypii* Glover on cucumber plants.Secondly,purely qualitative classes,based on size and number of insect patches or on the percentage of contaminated shoots,may be used.Such classes are generally used for large sampling units such as trees.

Finally,there are intermediate systems which are based on the number of sub-units (*e.g.*leaves of a plant) in each class of a set of qualitative classes.

This system was used by Lapchin to evaluate non-mummified *A.gossypii* on cucumber plants and by Boll for *A.gossypii* in open-field melon crops.A visual class system must be both simple and complete.Ease of use depends on the number of classes and therefore there should be a sufficient number to describe accurately the trends of variability in insect density,but not so high as to be difficult to remember.An optimal class number is generally between five and eight.Another condition required for the use of visual classes is that there must be a biological basis for their definition.

To be representative,a qualitative class set must cover the different kinds of patchiness of the species which may be encountered in the field.For example,on cucumber plants isolated colonies of *A.gossypii* a few centimetres in diameter will first develop around winged immigrants (slightly infested leaves).After several days of development,the colonies suddenly spread all over the leaf area (heavily infested leaves).These simple characteristics,which are associated with the size of the leaves that are heavily infested,define the classes of abundance.

Simplicity of the class system determines both the robustness of the results and the time required for field observations.In the example cited above,the observation of one sampling unit takes approximately 30 sec for each cucumber plant.The class system must cover the whole range of insect densities per sampling unit which may be encountered within the observation period and under different conditions.Thus,this range must be evaluated either from previous knowledge or from trials prior to defining the classes.

Calibration of Visual Abundance Classes

The results of visual observations may often be used without any reference to the number of individuals that they represent and,as such,these ranked qualitative data may be analysed using a large set of non-parametric statistic tools.This method has been used,for example,to evaluate the efficiency of biological control of the rose aphid on rose bushes in public gardens.

When more precise data are needed,each visual class must be calibrated by computing the mean and variability of the number of individuals actually present in the sampling units.This step is very timeconsuming because a large set of precise counts must be gathered so as to represent accurately the variability of the situations in which a given class may be chosen.

Classes Using Environmental Descriptive Variables

Calibration of the classes may be viewed as a statistical model having as a response variable the density of aphids,and as a categorical explanatory variable the visual abundance classes.A complex multivariate regression method,"projection pursuit regression",was adapted to this statistical data.Predictions of these models may be further improved by complementary explanatory variables.

This work,including calibration with complementary variables,was performed with four different class systems which were used to evaluate the density of the aphid *A.gossypii* and its parasitoid *Lysiphlebus testaceipes* (Cresson) on cucumber plants in greenhouses.

Two visual methods were used to estimate densities: "the detailed visual method" (DVM) for a leaf,and the "quick visual method" (QVM) for the whole plant.The class sets of DVM and QVM were built according to the apparent numbers of individuals in the observed sampling units.When the QVM was applied to healthy aphids,the four classes were based on the proportion of the area of leaf infested and on the size of the leaves.Precise counts were also made on the same sampling units and used as the response variable,and the data were divided into reference and validation sets.The reference sets were used to develop the regression models,and the validation sets to test their robustness.

The choice of complementary explanatory variables was crucial to the development of these regression models.These variables were selected for their influence on the goodness of fit of the models as well as for the time required for collecting.

For example,when using the detailed visual method and when the target population of the model was either healthy or mummified aphids,seven explanatory variables were used:

- The visual class of the leaf;
- The visual class of the non-target population on the leaf;

- The visual classes of the target population on the upper and lower neighbouring leaves on the same plant;
- The vertical rank of the leaf on the plant;
- The number of leaves on the plant;
- The number of leaves infested by the target aphid on the plant.

Such data can be easily gathered during sampling without significant additional cost.QVM sampling of whole plants yielded a mean error of approximately one class per plant (the limits of each class are in the ratio of).The DVM had a mean error of less than one class.The range of the residuals was generally the same for both the reference and the validation data sets,confirming the robustness of these models.

The same method has now been used to calibrate visual class systems for different pest species on vegetable crops in greenhouses (thrips on cucumber plants,aphids on tomato,melon,eggplant and sprout).Each time that a new regression model is tested,particular attention must be given to the development and sampling of the reference data sets,*i.e.*they must include the same combinations of variables which will be used in further field sampling.

TIME SPENT ON EVALUATING INSECT DENSITIES

Reducing the time spent on evaluating insect densities in sampling units has a cost,which is a decrease in the precision of the estimation.However,the time saved allows the observer to increase the number of units taken into account in a sampling plan,and thus to increase the precision of the mean and variance estimates of the density.

The gain in time is greatly increased when visual methods are used (a ratio of 1:10,when compared with precise counts).Mechanical methods,such as washing (ratio of 1:2),are much slower.However,the evaluation of the precision of visual methods requires a time-consuming calibration of the classes,which must be repeated for each study that deals with a different species,a different environment of the insects or a different scale of observation (*i.e.*plant or leaf).The decision to undertake such work will depend on the chance of building a "good" visual system.

We can summarize the four criteria of this evaluation as follows:

1. Insect density must be highly variable from one sampling unit (often a host plant) to another,and from one sampling date to another (this condition is easily met for numerous phytophagous insects whose densities vary on logarithmic scales from one host plant to another);
2. Most insects must be visible (for instance,such methods cannot be used for certain aphid species mainly located inside rolled leaves);
3. The visual classes must be simple and distinct,*i.e.*their boundaries have to be easily recognizable in the field;
4. These boundaries must be stable in time and space (for example,independent of the host plant growth stage).

The benefit of building such calibrated visual scales depends on which species is being observed and its environment.The scales are particularly useful for most aphid species,as they can reach very high densities and have strongly aggregated distribution patterns.Since the sampling units are not destroyed,crops can be easily monitored and the population dynamics studied separately in different fields.

Such a method permits large-scale surveys.An example is given in figure a set of cucumber greenhouses in Provence (France) was sampled weekly by using visual abundance classes and number modelling.A regular sampling grid was used in every greenhouse and required less than one hour of observation by two people on each sampling occasion.

5

The Role of The Home Garden

OBJECTIVE

By the end of this session,field workers will be able to understand the importance of the home garden in the daily lives of households.

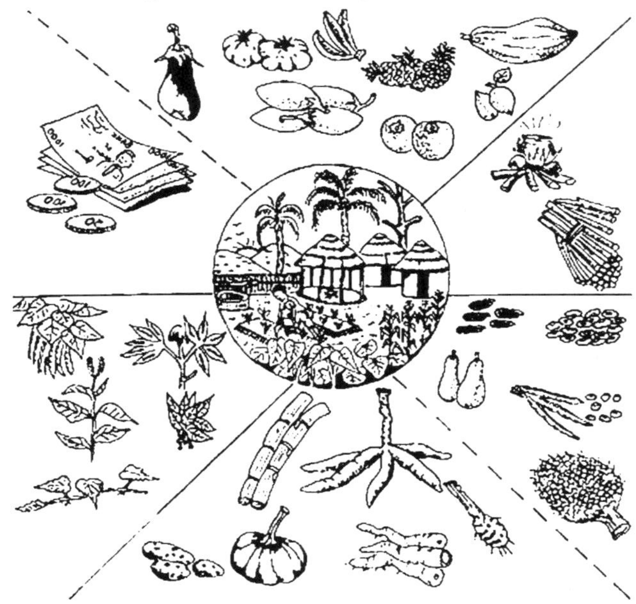

Fig. The home garden produces many different things: food,income,fuel for cooking,herbs,spices and flowers

OVERVIEW

The home garden is an extremely important piece of land in Africa.As its name suggests,it can be located either around or near the home,but frequently it is located close or adjacent to a permanent source of water,such as a river,pond or swamp.The home garden is often one of several field systems operated by a rural household.In peri-urban and urban areas,or in places where land is scarce,the home garden may be the only cultivated plot. African home gardens vary,depending on agro-ecological,climatic and cultural factors.In the humid and subhumid areas of central and West Africa,a home garden often includes a permanent agricultural plot or forest garden that contains perennial and annual crops.In some cases,the garden surrounds the homestead and paths lead to other field systems and production units devoted to annual crops for the market and home consumption.In other cases,the garden is located at a distance from the homestead,but close to a water source.

The biological diversity and complexity of home gardens decline with the transition from humid areas,where annual rainfall exceeds 1 100 mm,to semi-arid and arid regions where annual rainfalls range from 500 to 900 mm,to 250 to 500 mm in the Sahelian zones.Insufficient water is a major constraint to successful gardening in dry areas,yet even in these areas,gardening is possible.Some crops can be grown in the dry periods through appropriate soil management,and with cheap and effective solutions for harvesting and storing water. The home garden visit discussed in this session will help participating field workers gain an understanding of:

- the home garden as the most direct way of providing daily food throughout the year,particularly vegetables and fruits,and its function as a safety net during the lean season;
- the average size of local home gardens,the variety of crops grown and seasonal variations;
- the contribution that a well-developed,as opposed to a not-well-developed,home garden can make to family food security.

ACTIVITIES

The trainer visits the community and,in consultation with community leaders and households,selects a number of households with well-developed gardens and an equal number with not-well-developed gardens.The number of home gardens selected will depend on the number of groups of field workers participating in the programme.

The trainer explains the purpose of the visit and makes appointments for home garden visits.Special care should be taken to include households that are headed by women.

*Discussion.*After introducing the aims of the session and defining the objectives for the first home garden visit,the trainer facilitates a discussion among the field workers on the following points:

- the definition of a home garden and its importance to households in the area;
- the role and use of the home garden;
- foods commonly produced in the home garden and seasonal variations.

The field workers then share their views and experiences on the topics.

*Planning the field visit.*The trainer presents the different planning stages in preparation for the fieldwork.These include:

- identifying objectives for the household assessment;
- identifying and notifying the communities to be visited;
- reviewing available data;
- deciding which additional data should be gathered;
- identifying which groups or individuals in the community are knowledgeable enough for group or individual interviews;
- drawing up a checklist of information for each interview;
- selecting assessment techniques for gathering needed information.

The trainer asks the participants to identify the objectives of their first field visit.This may include identifying:

- the criteria of a well-developed or a not-well-developed home garden;
- which households have well-developed home gardens and which do not;
- the types of crops grown in the home garden and their uses (e.g.home consumption,sale);
- which households are not able to meet their nutritional needs.

Once the participants have reached an agreement on the objectives of the field visit,they need to identify what information they need and how they will gather this information according to the above planning guide.During the first field visit,participants will conduct key informant (village representatives) and household (husband,wife,children) interviews.

*Interacting with the community.*After participants introduce themselves and explain the purpose of their visit,a good starting point is to ask about the farmer's present activities.Other good introductory questions include ones about general aspects of the community,such as the number of households in the village,existing services and infrastructure,and main crops grown.Field workers must determine the mood of the discussion before asking more sensitive questions,such as those concerning the types of food eaten,the number of meals eaten daily,and whether or not enough food is produced for household consumption needs.

*Field workers' attitudes towards the community.*Before field workers can assist a community,they must give local people the opportunity to describe how they do things,what they know and what they want.Therefore,field workers should strive to understand the local situation,and facilitate and support communities' own efforts to find and implement solutions.As outsiders,field workers are listeners and facilitators rather than teachers and experts.They observe,ask questions,listen and gently probe issues requiring clarification.In doing so,they can provide local people with the opportunity to analyse their own problems.This respectful stance generally leads to improved rapport and collaboration between community members and field workers.In this way,field workers and community members can combine knowledge and expertise,and work together to find solutions.

*Group preparation for the fieldwork.*There will be three field visits during this training course.Each group of field workers should appoint its facilitator who can introduce the group to the community and to households,explain the purpose of the visit and open communication with household members.Other group members will facilitate when issues arise in their area of expertise.The group also selects a recorder,who should carefully write down all responses by community members during the discussion.To give everyone a chance to practise the different roles of facilitation and recording,the tasks can be rotated among team members during the three field visits.

*Organizational aspects of the fieldwork.*The group starts the field visit by calling on the local leaders and meeting with a group of community members before visiting areas of interest and households.

*First home garden visit.*The first home garden visit gives the participants a basic understanding of the structure,functions,limitations and potentials of home gardens in the area.After the trainer has covered the preparatory and organizational aspects of the field visit,the group proceeds to the community meeting point,where the group facilitator introduces the team and explains the purpose of the visit. The field workers begin by asking general questions of village representatives before proceeding to specific home gardens.As they walk through the village and the home gardens,the field workers can discuss with the villagers what they are seeing (e.g.soil,rivers,crops,housing,water wells,sanitary facilities,health centres and schools).This is called a *transect walk* and can provide field workers with a cross-section of the area.A transect walk can also be repeated with different community members,such as groups of men,or of village elders or of young people,to reflect upon gender- or age-based differences in activities or access to resources.

*Small group discussion.*Participants return to the training venue to summarize their findings,organizing the information by type of household (i.e.those with well-developed gardens and those with not-well-developed gardens).

*Whole group discussion.*Each group presents its findings in plenary followed by discussions.The trainer asks groups to compare the key differences observed between the well-developed and the not-well-developed home gardens,and encourages field workers to discuss the contribution that home gardening makes to family food supplies in general and to specific families' food supplies in particular.At the end of the session,the trainer summarizes the findings.

THE HOME GARDEN IS AN IMPORTANT LAND UNIT

In many humid and subhumid areas of Africa,home gardens are a common feature of the food production system.Because of lack of water,they are less frequently cultivated in Africa's semi-arid and arid zones.

In most African communities,people rely on one or two staple crops such as maize,millet,sorghum,rice,teff,cassava,yam,sweet potato,plantain and enset.These crops tend to provide the bulk (about 60 to 80 percent) of the energy intake of household members.Home garden foods supplement staple crops and add variety and nutritional value to the diet.They typically include roots and tubers,green leafy vegetables,legumes and fruits,which are rich in micronutrients - such as vitamins A and C,iron and,sometimes,vitamin B - and in some cases contain appreciable amounts of protein and oil/fat.

Home garden foods also have an important "safety net" function.During the lean season,when the staple foods have been depleted and before the new harvest is ready,home garden foods can augment or replenish family food supplies.Unlike field crops,home garden foods can be cultivated and made available for family consumption year round,if there is enough water available.

A WELL-DEVELOPED HOME GARDEN HAS MULTIPLE FUNCTIONS

Well-developed home gardens can be found on walks through most villages.Local households with well-developed home gardens possess the ideas,skills and resources to produce a variety of crops,including some staples,roots and tubers,legumes,vegetables and fruits,as well as livestock and,sometimes,fish.They know how to layer plants,combining tall plants with shorter plants,as well as plants that mature at different times.Their animals consume plants from the home garden and return nutrients to the soil in their manure.In addition to producing food,these gardens provide income from the sale of produce,inputs for farm development activities,and non-food items including spices,herbs and medicinal plants,plus they provide space for processing,preparation and storage of food.

*Income from the sale of home garden produce.*The sale of home garden produce can make a substantial contribution to household income,especially during seasons when other sources of employment and income are limited or harvests are reduced by natural disasters (e.g.floods,outbreaks of pests or animal disease,or personal illnesses).At these times,income from a home garden can be used to purchase food items that the family cannot produce,thus adding variety to meals and supplementing production.Home garden income can pay for daily essentials and services,such as soap,clothes,school fees,medicines and farm inputs that cannot be produced by the household.

*Inputs for farm development activities.*Important farm development activities take place in home gardens and generate inputs for other farming activities.For example,a home garden can produce planting materials,such as sweet potato vines,cassava cuttings,fruit-tree seedlings and vegetable seeds.When animals are part of a home garden (e.g.sheep,goats,poultry and perhaps cattle and pigs),they supply food,income,manure and draught power.The animals,in turn,benefit from unused plant residues,such as the stalks of cereals,vegetables and legumes.The home garden is also a place for experimenting with new crops and farming techniques.

Non-food items. Where climatic patterns allow and where they are culturally relevant,non-food items such as medicinal herbs,spices and flowers are produced.Home garden trees and shrubs provide shade and act as natural windbreaks and barriers to scavenging animals,but they also can produce fencing and building materials,and fuelwood.

Processing and storage areas. The home garden is also a place where processing and storage activities may occur.In order to have a stable,year-round supply of cereal and legume crops from the field,households need adequate storage and processing facilities.Storage structures for cereals and legumes are often located within the homestead and must be appropriately constructed in order to prevent food losses caused by insects,rats and other pests,and nutritional deterioration resulting from fungi and food rot.

Fig. frican home garden

PRACTICAL NUTRITION FOR FIELD WORKERS

OBJECTIVE

By the end of this session,field workers will appreciate the importance of adequate household food supplies as a basic condition for the nutritional well-being and health of all family members.They will:

- understand the meaning of the term "nutrition";
- know the differences in the composition of foods commonly eaten;
- know the differences in the nutritional requirements of different family members;
- understand the effects of inadequate food intake on nutritional status and health.

Fig. A family's nutritional well-being depends on access to a nutritious diet at all times

OVERVIEW

Most people eat because they feel hungry.Although the sensation of hunger tells them to eat,it does not tell them *what* to eat.Field workers promoting home gardening need to have a basic understanding of nutrition,in order to help household members grow and select a variety of foods,develop healthy eating habits and lead active and healthy lives.

Nutritional well-being requires access to enough nutritious and safe food to meet the dietary needs of all members of the household throughout the year.Food security and nutritional well-being mean more than just producing enough food.People must also have *knowledge* about nutrition,especially of what kinds of foods to eat and how to prepare them in the right quantities and combinations,and in a way that is safe and clean. It is important for mothers to know about proper feeding practices and to have enough time for food preparation and frequent child feeding.In addition,access to sufficient clean water and a healthy environment mean less risk of infectious diseases.A healthy body is better able to utilize foods,because illness can interfere with nutrient absorption,especially such illnesses as diarrhoea,parasitic infections and measles. This session defines the meaning of *nutrition* and identifies the main nutrients in commonly eaten foods.It discusses the role and importance of nutrients in different foods,the nutritional needs of different household members and what happens to the body if people do not follow a healthful diet,practise adequate personal and food hygiene or have access to adequate and safe drinking-water or health care.

ACTIVITIES

Discussion. The trainer states the objective of the session,and then,using the appropriate technical notes,initiates a discussion based on the following questions:

- What is nutrition about?
- What are the nutrients in foods and what are they for?
- Do the nutritional needs of different household members vary,and if so,how and why?
- What happens if the nutritional needs of different household members are not met?

Discussion of the value of different foods. Keeping in mind the home garden crops identified during the first household visit and other foods grown on the family farm,participants should discuss the importance of each crop and how it is used in meal preparation.They should then classify the foods by their key nutrient(s),for example,energy (carbohydrates and fat),protein,vitamins and minerals.After this discussion and classification,the participants should compare their knowledge of food nutrients with the information.

Discussion of dietary deficiencies. The trainer reproduces,"Symptoms of nutritional deficiencies",on the flip chart and invites the participants to discuss,based on their knowledge and experience,what happens to the body when a person does not get enough energy producing food,or when a specific nutrient is lacking or inadequate in the diet.

The field workers then read Information Sheet 2,"Food and nutrition problems".Together they adjust their responses accordingly,and the trainer answers any questions that arise.The trainer then conducts a practical session on assessing child nutritional status using measures of mid-upper-arm circumference (MUAC).Prior to this session the trainer should invite several mothers with their children (aged one to five years) to be present.Participants practise taking mid-upper-arm-circumference measurements on the children using a MUAC strip or a simple tape measure.

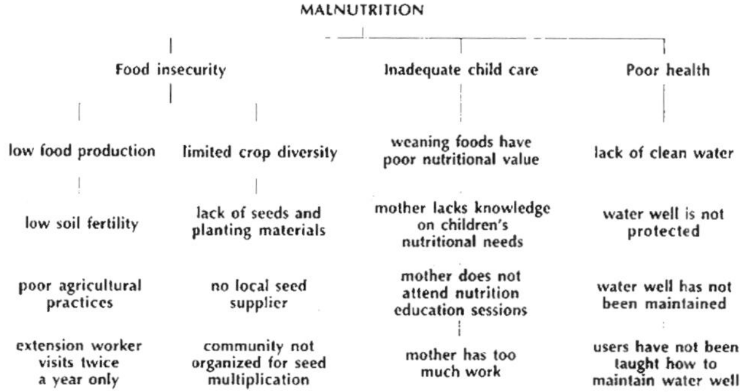

Fig. Causal model of malnutrition

Exercise. After having reviewed the signs and symptoms of malnutrition and micronutrient deficiencies,the trainer initiates a discussion of the causes of malnutrition and other nutritional deficiencies that are present in the community.The trainer facilitates this session and helps the participants build a causal model or problem tree (a diagram of the causes of malnutrition and how they are interrelated) that focuses on what they consider to be the causes of malnutrition and nutritional deficiencies in the communities they are familiar with.Creating a causal model helps field workers better understand the interactions of the different factors that influence nutritional outcomes.

Nutrition is about food and how it is used in the body

Nutrition is an area of knowledge and practice concerned with food and how it is used in the body.This includes how food is produced, collected, bought, processed, sold, prepared, shared and eaten,as well as how food is digested,absorbed and used by the body,and how it finally influences well-being.

If eaten in the right amounts and combinations,food provides all the nutrients the body needs for proper growth,functioning and mental development,and to keep people healthy.

Food is made up of a combination of nutrients

Food is made up of nutrients such as carbohydrates,proteins,fats,vitamins and minerals.Vitamins and minerals are called micronutrients.They are needed in much smaller amounts than protein,fat and carbohydrates but are essential for good nutrition.A healthful diet is one that supplies adequate amounts of all these nutrients.

All foods from plants and animals contain a mixture of nutrients.Each food,however,has a different amount of each nutrient.Except for breastmilk in the first six months of life,no single food provides all the required nutrients.

To achieve a balanced diet and stay healthy,people must eat a variety of foods each day.For example,staple foods such as cassava,yams,taro (cocoyams),sweet potatoes and plantains are good sources of carbohydrates,but contain relatively little protein.Cereals,including maize,sorghum,millet and rice,in addition to being rich in carbohydrates,also contain significant amounts of protein and B vitamins.

Legumes such as cowpeas,pigeon peas, soybeans, beans, groundnuts, bambara groundnuts and horsebeans are rich sources of protein and iron,while soybeans and groundnuts are also good sources of fat. Roots and tubers,legume seeds and cereals are relatively poor sources of vitamins A and C,while fruits and vegetables,including cassava,cowpea,bean and sweet potato leaves,contain significant amounts of vitamins and minerals.

To balance their diets,people must complement staple foods with legumes or foods from animal sources that are rich in protein and oil or fat; vegetables such as pumpkin leaves or amaranth; and orange- and yellow-coloured fruits and vegetables such as mangoes,papayas and pumpkin,which are extremely rich in vitamin A.Information Sheet 3,"Recipes for nutritious dishes",provides a family mixed meal guide,which gives suggestions for the alternative types of food mixtures that constitute a nutritionally adequate meal.

The nutrient content of food changes with the degrees of freshness of the food and the methods of processing,preserving and preparing it.For instance,if the outer skin of cereal grain is removed during milling,most of the cereal's minerals and vitamins are lost,and the vitamin C content of vegetables diminishes if the vegetables are overcooked.The nutritional values of some of the garden and field crops commonly grown in different parts of Africa,including processed foods such as sugar,oils and maize flour.

Nutrients are needed to keep the body active and healthy

Most farmers know from experience that crops need certain nutrients in order to grow well.Plants get these nutrients from the soil or from fertilizer.In a similar way,people require certain types and quantities of nutrients from their diets,from the time a baby is conceived in the mother's womb and throughout life. Each type of nutrient serves particular functions: carbohydrates and fats are the main sources of food energy.About 50 percent of the carbohydrates and fats in the body are "burned",or broken down,to produce the energy the body needs to perform different physical activities; the rest is used by the body for growth and general maintenance and in the renewal of its tissues.

Fat is a particularly concentrated source of energy and contains twice as many kilocalories as carbohydrates and proteins.Fats,in addition to providing energy,are also needed for the absorption and use of some vitamins,especially vitamin A.Proteins are needed to build and maintain muscle,blood,skin,bones and other tissues - they are the primary building blocks of the body.Protein is especially important for children,and pregnant and lactating women.

Vitamins and minerals help the body work properly and stay healthy.They contribute to tissue repair,and ensure children's healthy growth,mental development and protection from infection.

Inadequate dietary intake in young children slows down or stops growth,weakens the body and impedes intellectual development.Children of school age who do not get enough nutritious food are unable to concentrate and do not learn as well as well-nourished children.

An inadequate diet can affect older children and adults of the household as well.Many households have limited labour available for work on farmland.A poor diet results in reduced work capacity and illness.Illness means visits to the health centre,which means lost working time and lost money.These losses can be reduced or prevented if everyone in the household eats enough healthful food,drinks clean water and practises good hygiene.

Inadequate dietary intake can also have far-reaching effects on the socio-economic development of a country.This is a situation that no country can afford.Actions are needed,therefore,that address the causes of inadequate dietary intake and improve the nutritional status and health of children and adults.

The amount of nutrients needed varies for each person and at different stages of life

The amount of energy and nutrients people require from their diets in order to stay healthy and active varies with age,sex,pregnancy and lactation,level of activity and state of health.

The most critical stage of human development is from conception to about three years of age,a period when physical growth occurs most rapidly.It is crucial,therefore,that small children and pregnant and lactating women in particular receive the right amount of nutritious food in order to ensure proper growth,brain development and resistance to infection of the foetus and young child.

Food given to children in the first year of life must have sufficient energy and nutrients to support rapid growth and development.A child under one year of age,for example,requires almost twice the amount of energy per kilogram of body weight as a teenager or adult.For example,a child less than one year needs 110 kcal/kg body weight per day; a child less than five years: 95 kcal/kg body weight per day; a 12-year-old male: 64 kcal/kg body weight per day; and a 16-year-old male: 44 kcal/kg body weight per day.

When breastmilk becomes inadequate to support healthy,normal growth,at about six months of age,children must be given foods that are nutrient dense (i.e.with high concentration of nutrients relative to the volume measure).This is particularly important because children at this age have extremely small stomachs and can take in only limited quantities of food at a time.Frequent feeding (four to five times a day) with foods,in addition to breastmilk,is therefore important to ensure that children get sufficient energy and nutrients to grow normally and stay healthy.Mothers should continue breastfeeding on demand until children are 18 months to two years of age.

Pregnant women need additional food of good quality to meet the nutrient needs of a growing foetus (baby in the womb).Lactating mothers require additional energy and nutrients to produce enough milk for a small child.

Note: In Africa the dietary intake of fat or oils is often low (i.e.less than 10 percent) because refined oils and animal foods that are sources of fat are expensive.

Everyone needs to eat some fat,however,because everyone,especially children,requires essential fatty acids.Also fat makes food less bulky and helps the body in its absorption of some vitamins,especially vitamin A.As a general rule,about 15 to 35 percent of kilocalories should be obtained from fat.Foods that are rich in fat include groundnuts and oil crops,such as sesame and sunflower seeds.Both coconut and red palm oil are excellent sources of essential fatty acids.Special efforts are needed to assist household to increase the production of oil- or fat-containing crops so as to increase their supply and enhance access at household level.

Table. Daily requirement of energy,protein,fat,vitamins A and C,and iron for different sex and age groups

Family member	Age	Energy (kcal)	Protein (g)	Fat (g)	Vitamin A RE (μg)	Vitamin C (mg)	Iron (mg)
Man (active)	18-60	2 944	57	83	600	45	27
Woman (child-bearing age)		2 140	48	59	600	45	59
Woman (pregnant)		2 240	55	65	800	55	c
Woman (lactating)		2 640	68	73	850	70	95
Child (1)	Under 1	800	12	*	375	25	19[b]
Child (2)	5	1 510	26	42	400	30	13
Child (3)	12**	2 170	50	60	600	40	29
Child (4)	14**	2 620	64	73	600	40	29

Note: 1 g protein or 1 g carbohydrates = 4 kcal; 1 g fat = 9 kcal; 1 g alcohol=7kcal.Fat requirements were calculated to provide 25% of average energy requirements. The assumption is that breastmilk satisfies this child's fat-intake requirements.

Data refer to male children.[a] The requirements are based on a low iron availability diet (i.e.5% ofironabsorbed).

[b] Bioavailability of iron during this period varies greatly.

[c] It is recommended that iron supplements be given to all pregnant women because of the difficulties in correctly evaluating iron status in pregnancy.

Daily household food requirements expressed in household measures

Table. Food quantities that meet daily energy and nutrient requirements ofdifferent household members

Family member	Maize flour		Beans		Cassava leaves		Cooking oil
	(g)	*Cups or local measure*	(g)	*Cups or local measure*	(g)	*Local measure*	(g)
Man (18-60 years old)	560		200		110		40
Woman (child-bearing age)	460		150		100		40
Woman (pregnant)	500		150		110		40
Woman (lactating)	500		200		160		40
Children 2-3 years	200		100		80		30
Children 5-6 years	250		150		100		30
Children 10-12 years	350		200		100		30
Children 14-16 years	400		225		120		30

Since the participants may not have access to scales,the daily food requirements make more sense if expressed in terms of food *quantities*,using local household measures.An example of the approximate food quantities (in grams) needed to construct a basic low-cost diet that meets the energy and nutrient needs of different household members.Only a few food items have been used to keep the diagram simple; however,people should eat a variety of foods,including fruits,every day.Trainers should convert these quantities into the local measures (e.g.cups,spoons,*mudus,olodo* or *tiers*) that are commonly used in a given home or at the local market.

In a cereal-based diet,about one-third of the required fat comes from cereals and the rest from other foods,whereas in a root-and-tuber-based diet,virtually all the fats come from relishes or snacks.

Based on information,a household comprising a father,mother (lactating) and four children aged 2,5,12 and 14 years would require about 2.3 kg maize flour,1.1 kg low-fat legumes,such as beans,and about 0.7 kg green leafy vegetables.During meal preparation,oil is often added,when available,to relishes or sauces to make them tasty.The quantities of oil vary depending on what the family can afford,but at least 200 g (1 cup) per day for cereal-based diets and 300 g (1.5 cups) per day for root-and-tuber-based diets are needed to meet nutritional needs,enhance the foods' taste and palatability (i.e.make eating more pleasurable),and make swallowing easier.Pounded groundnuts,melon seeds and sesame seeds are often used as alternatives and added to vegetable relishes or soups,making these dishes tastier.Energy-rich snacks for children complement main meals and add energy to the diet.Fruits,which are rarely eaten as part of a meal,should be included,especially if the diet lacks micronutrients.

Table. Home garden crops that are rich in key nutrients

Energy	Protein	Fat	Vitamin A	Vitamin C	Iron
Avocado	Bambara groundnut	Avocado	Amaranth or African spinach	Baobab fruit	Beans/peas*
Bambara groundnut	Baobab seed	Bambara groundnut	Bean leaves	Cabbage	Kidney
Banana/plantain	Beans/peas	Butter	Bitter leaf	Cashew nut	Liver
Barley	Cowpea	Coconut cream	Carrot	Citrus	Meat/chicken, fish
Breadfruit	Eggs/milk/cheese	Groundnut	Cassava leaves	Custard apple	Some green leafy vegetables*
Cashew nut	Groundnut	Oil from groundnut, maize, nug, safflower, sesame, soybean, sunflower or any other oilseed	Cat's whiskers(*Cleome gynandra*)	Guava	
Cassava	Kapok seed	Shea butter nut	Chillies	Mango	
Coconut	Meat/chicken, fish	Some insects and caterpillars	Jute	Papaya (ripe)	
Enset	Melon and pumpkin seeds	Soybean	Kale	Passionfruit	
Groundnut	Oyster nut		Liver	Pineapple	
Maize	Pigeon pea		Maize (yellow)	Sweet pepper (capsicum, if orange)	
Millet	Some insects and caterpillars		Mango (ripe)	Sweet potato (yellow- or orange-coloured)	
Oil from groundnut, maize, nug, safflower, soybean, sunflower or other oilseeds	Soybean		Okra	Tomato	
Oyster nut			Papaya (ripe)		
Rice			Pumpkin		
Shea butter nut			Rape		
Sorghum			Red palm oil (unrefined)		
Sweet potato			Rosella		
Taro			Sweet potato leaves		
Teff			Sweet potato (yellow- or orange-coloured)		
Wheat					
Yam					

Depending on the usual meal frequency,the recommended quantities can be spread out over two or three meals per day for adults.Because of their small stomach size,children under five years of age need more meals and should be given snacks between the main family meals.These snacks can be in the form of leftovers from the main meals.

A practical approach to healthful eating is to ensure the consumption of a variety of foods,focusing on the staple foods,as well as those that are ingredients for relishes,West African soups or stews.The relish,soup or stew often provide additional energy (from beans,peas,groundnuts and oil),additional protein (from beans,peas,groundnuts,meat and fish) and essential micronutrients (from vegetables,fruits,meat and fish). · Absorption of iron in these foods is increased by combining them with vitamin- C- rich foods,for example,by eating an orange or guava at the end of a meal.

Fig. Home garden produce

OVERVIEW

An adequate diet must contain enough food in the right combinations for everyone,young or old,male or female.Different nationalities,groups of people and cultures have their own food habits,traditions and preferences that influence food choice,cooking methods and diet composition.Factors such as food cost and seasonal availability must also be considered when discussing food and nutritional patterns and ways of improving food choice and meal planning and preparation.During this session,participants will use their knowledge of the local culture,food production,home gardening and food habits to evaluate the nutritional adequacy of meals.They will also learn how to help people improve the nutritional value of local diets and how to encourage improvements in crop choice,land use and other home garden practices that contribute to the nutritional well-being of the community.

ACTIVITIES

Discussion of nutritionally adequate meals. After outlining the objectives of the session and explaining the activities to be undertaken,the trainer asks the participants to use the flip chart to present the types of dishes that are prepared locally,starting with main dishes.The participants then list the dishes (including their ingredients) that accompany the given main dishes (e.g.the relish).Main dishes made from a mixture of food items and considered "complete" meals needing no accompaniment or relish (e.g.West African soup or stew) should be listed separately. The trainer then introduces the "Family mixed meal guide 1" from Information Sheet 3 and explains how it is used to plan and prepare family meals that are healthful and nutritious.Special attention should be given to essential foods and nutrients that may be lacking or in short supply due to seasonal variations in the locality.Keeping the family mixed meal guide in mind,the members of the group should decide which of the commonly consumed meals they present are nutritionally adequate and which lack essential food items,and then explain why.For those meals that are nutritionally inadequate,the field workers should be able to suggest different ways to improve them.

The participants should consider the following questions:

- Which food items contain more vitamin A: 2 cups amaranth,3 cups cowpea leaves or 4 cups okra pods?
- What are the major nutrients in palm oil or groundnut oil?
- What are the price differences among similar foods (e.g.different types of leafy vegetables,different types of legumes,different types of oils)?
- Which specific foods do you consider "good buys" for households,and why?

Note: Only 1.5 cups chopped dark green leafy vegetables,or a medium-size yellow-coloured sweet potato,or 2 teaspoons unrefined red palm oil (10 ml) or one small carrot a day will provide enough vitamin A for a child to avoid eye damage and reduce his or her risk of getting measles and other diseases.Adding a small spoonful of oil to a child's food enhances vitamin A absorption by the body.

Exercise. The trainer draws a table listing the main local foods,such as cereals,roots and tubers,oilseeds and pulses,fats and oils,fruits and vegetables,and then asks the field workers to indicate whether the nutritional values (i.e.energy and nutrient content) of each of the local foods is good (+),fair (0) or poor (-).

Snacks. The trainer defines the term "snack" (a food that is eaten between main meals) and then discusses snacks and the contribution they make to the diet,particularly to a child's diet.The trainer refers the field workers to Information Sheet 4,"Nutritious and tasty snacks for young children",as reading material.

The group makes a list of local snack foods,estimates the nutritional value for each snack food and indicates which are particularly rich in certain nutrients and how they contribute to improving daily nutrient intake.

Planning daily family meals. Using the Technical Notes section in Session 3 as a guide,the trainer initiates a discussion of why people prefer different foods (i.e.why they have food preferences).Then the participants arrange themselves in small groups and review the daily meal frequency of the local community,any seasonal variations that exist and the local conditions and factors (e.g.the mothers' many responsibilities,the fathers' working away from home,lack of fuelwood) that determine such meal patterns.They also discuss how seasonal variations in food availability affect the main dishes,the relish ingredients and the snacks given to the children. Information on the size of the family and the age and sex of its members has already.The number of family meals will depend on the outcome of earlier discussions on daily meal frequency,with two to three family meals being considered as average.Snacks for young children may be included and should be indicated.The meals may consist of *some* or *all* of the following:

- breakfast
- morning snack (for young children)
- lunch
- afternoon snack (for young children)
- dinner

During the exercise,the group members should aim at getting the best nutritional benefits as economically as possible,taking into account the current prices of food commodities.Keeping these things in mind,they should:

- plan meals using the foods either grown in the home gardens they visited or referred to by household members during the first visit;
- plan meals with variety and that are nutritionally balanced;
- specify quantities of the required food items,including those that may have to be purchased;
- estimate the cost of the day's meals,including food items that come from the family garden or farm.

Information Sheet 3,"Recipes for nutritious dishes",may provide the field workers with helpful hints.

Each small group should present its example to the whole group and respond to comments.The whole group should note the most common meal suggestions and divide the meal components into:

- food items that can be grown in the home garden or on the farm;
- items that must be bought from the market or at a shop.

THE NUTRITIONAL VALUE OF A FOOD

Agricultural extension workers often think of food in terms of different crops or animals.Some food crops are difficult to grow while others are relatively easy to grow.Some bring good prices at the market while others do not.Home economists think of the convenience of processing and preparing certain foods,while nutritionists and public health specialists think of the nutritional value of different foods and their effect on health in certain quantities and combinations (e.g.in preventing *kwashiorkor*,*marasmus*,night blindness,allergies,diabetes and obesity).

Families,however,consider many other things.They may choose the food they eat for a variety of reasons,including habit; tradition; convenience in processing and preparation; a preference for the food's taste,texture and colour; the food's availability depending,for example,on season,adequate storage and consumer demand; its price and its suitability for an occasion (e.g.a wedding,funeral,childbirth).Often people,particularly the elderly,feel they have not eaten properly if they have not consumed the staple,or any other familiar food,at least once a day.Field workers must take all these factors into account when exploring possible ways to assist communities in improving their eating habits and nutritional status.

SNACKS FROM THE HOME GARDEN ARE AN IMPORTANT SOURCE OF NUTRIENTS

Snacks are eaten between meals and are often a normal part of a child's diet.They are usually foods that provide energy quickly and are eaten raw (e.g.fruits and sugar cane) or cooked.Snack foods consisting of beans and nuts can provide significant amounts of protein and fat,and those prepared from,fruits and vegetables provide important vitamins and some minerals.

Snacks can be seasonal foods,particularly in rural areas of the subhumid and semi-arid parts of Africa.Street food vendors in some parts of Africa often specialize in certain snacks (e.g.bean cakes,boiled or roasted cassava and roasted groundnuts,oilseeds or grain legumes).

Depending on the region,good snacks may include:

- cooked or roasted roots,tubers or cereals (e.g.yams,sweet potatoes or maize);
- starchy fruits (e.g.bananas or cooked breadfruit);
- fruits (e.g.mangoes,guavas,papayas or a wide variety of wild fruits);
- sugar cane or sweet cereal stalks;
- roasted or fresh nuts,legumes or seeds (e.g.groundnuts,oyster nuts,chickpeas or bambara groundnuts);
- cowpea or melon seed fried cakes (e.g.*akara* or *akara egusi*).

A home garden containing a variety of crops can provide many snacks that supply essential nutrients and contribute to the nutrition and health of all family members,particularly children.

Nutritionally balanced and tasty meals can be planned by adding nutrient-rich food items to the staple food

Family meals. Locally available staples generally form the basis of a meal,but a meal becomes nutritionally adequate and tasty only if a relish or soup (consisting of beans or groundnuts,vegetables,fats or oils,condiments and spices) and fruits are eaten with the staple.The type of relish or soup that accompanies a main dish,and the ingredients used to prepare it determine the nutritional adequacy of the meal.

Apart from animal products,most of the ingredients used to prepare a relish,West African soup or other accompaniments to a main dish come from the home garden.Where foods of animal origin (meat or fish) are not readily available,the quality of the diet can be improved by providing a variety of vegetable products at each meal (e.g.beans,lentils or groundnuts with green leafy vegetables).

Some meals are adequate from the point of view of quality while others are not.The family mixed meal guide in Information Sheet 3 provides a quick and practical way to evaluate the quality and variety of food items in a meal.

The information provided offers a practical way to plan meals and determine which food quantities meet the nutrient needs of different members of the family.

*Meals for infants and young children.*Starting at two years of age,children may eat foods from the family pot,but children under two years of age require special food.Their teeth are usually not developed adequately to chew tough foods,and they must adjust gradually from breastmilk to semi-solid foods,and ultimately,the food from the family pot.

During this transition period,the texture and thickness of a child's food has to be appropriate to the level of development of the child,and the food's energy and nutrient content must support rapid growth.It is also important that the processing and cooking methods required for a child's diet do not place an extra burden on the mother.

Information Sheet 5 provides useful information on weaning foods and gives some recipes for energy-dense infant feeding mixtures prepared in different parts of Africa,using locally grown foods.

FACTORS AFFECTING HOUSEHOLD FOOD SECURITY AND NUTRITIONAL STATUS

OBJECTIVE

Fig. Food groups

By the end of this session,the field workers will be able to:

- understand the local food system and the different factors that contribute to household food security;
- understand the interactions between household food security,health and care in determining the nutritional well-being of household members.

OVERVIEW

Having sufficient good-quality food available for household consumption depends on several factors.Household production is one way to ensure that there is food within the home at all times,but producing enough of a variety of foods in the home garden and in the fields requires access to adequate resources,including land,water,seeds,tools,knowledge,skills and labour.Roads and transport to markets are necessary for buying and selling food and other essentials.Household members also need access to other commercial and government services,and off-farm employment during periods of low agricultural labour demand.

In rural areas with limited income-earning opportunities,the ability to produce most foods in the home garden and on the farm,without depending on market purchases,means a better guarantee of household food security.Household food security depends on a regular and sustainable supply of food throughout the year.In many rural areas,however,households often face food shortages because crop production is seasonal and at times inadequate.This session reviews the factors that contribute to food security at the household level,and their impact on dietary intake and nutritional status.The causal model constructed during Session 2 is used to explore the contribution that other factors,such as adequate child care,health and sanitation,make in determining nutritional outcomes.

At the end of this session,the participants will analyse further the problems and constraints that exist in the food system,starting with food production and ending with dietary intake and utilization of the food ingested.Drawing upon their knowledge of and experience with local communities,group members will explore the opportunities that exist for enabling households to address some of the identified constraints.

ACTIVITIES

*Discussion.*After presenting the objectives of this session,the trainer reviews the causal model of malnutrition that the field workers constructed,and then asks the field workers to think about all the factors that contribute to adequate,year-round access to food at household level.Before starting small group discussions,the trainer introduces the concept of the local food system in plenary,requesting the participants to list all the stages in the food system,from production to consumption.The small group discussion that follows should focus on the following questions:

- What factors determine food security at the household level?
- What are the stages in the food system?
- What problems and constraints do local communities or individual households experience at each stage?
- What factors influence nutritional status and food choice in the local communities?
- What can individuals or the community do to address some of the identified problems and constraints?
- What are the positive nutritional and home garden practices and opportunities that can help people make improvements?
- How can a field worker help a community consider such opportunities?

*Small group work.*The participants divide into groups and each group considers the above issues.

*Presentation of results.*Each group presents the outcome of its discussions in plenary.The whole group subsequently draws up a final list of problems and constraints,as well as the opportunities that exist in the local food system for improving household food security.A flip chart or any other media available to the participants may be used during the presentation.

*Summary of group presentations.*The trainer summarizes the main findings and conclusions of the discussion.These will serve as background information for Sessions 7 and 8.

*Summary of session.*On the basis of what the field workers have learned,the trainer concludes the session by summarizing the interrelationships of the various factors affecting household food security and nutritional outcomes.The trainer also discusses the roles that knowledge,education,and women's and men's different roles and responsibilities,and their access to and use of resources,play in overall nutritional improvement.

Household food security,care and health are essential elements for ensuring good nutrition of all household members

Household food security is defined in Session 1 as "access by all people at all times to the food needed for a healthy life".Households can obtain food supplies either from their own food production or from food purchases,but more often it is through a combination of both.Some factors that help assure that communities have enough of a variety of foods at the household level are access to sufficient water,fertile land,seeds,planting materials,agricultural implements,extension advice,credit,good storage and a sufficient number of family members who are healthy and strong enough to work on the farm and undertake off-farm employment.

Many rural communities,however,do not have year-round access to adequate amounts of either fresh or processed staple foods,and their access to fresh vegetables and fruits tends to be seasonal.Frequently,households sell too much food either because they need cash,or facilities for storing and preserving foods for longer periods are poorly developed or non-existent.

Households can use several strategies for ensuring continuous access to a variety of nutritious foods.These include the year-round production of a variety of home garden foods (given that some water is available) and the preservation,processing and adequate storage of foods.Households can thus insure themselves against seasonal shortfalls,but it requires planning ahead and thinking about what is the most effective use of available resources so as to prevent hungry periods.Appropriate technology options are available to extend garden production and enable households to process and store perishable staples,legumes,vegetables and fruits,in order to extend their availability and enhance their marketing potential.

Care. Household food security is a precondition for adequate dietary intake,but it does not always translate into good nutritional status.The provision of adequate care especially for small children and other nutritionally vulnerable household members is crucial to achieving good nutritional status.Care includes such things as having knowledge of the differing food and nutritional needs of household members.Such knowledge helps the caretaker,usually the mother,make the right decisions on what foods and how much to prepare and serve during meal times to the children and adults.Apart from having nutritional knowledge and skills,the caretaker must also have adequate time for child care,food preparation and child feeding.

In many African cultures,women are the main household food producers and caregivers.They also have many other responsibilities,such as fetching water and fuelwood,and storing,processing and marketing food.During peak seasons of agricultural activity,the danger is particularly high that children's needs for care,proper food and frequent feeding will not be met adequately.Despite the many constraints,there are many options to enable households to better care for their vulnerable members.These include the use of available technology that relieves demands on women's time,such as providing better access to water.Other options are more people-centred and require the community or individuals to change established traditions or behavioural patterns.This means that all community members,adults and young people,should become better educated as to the advantages of good nutrition and health,as well as the many possibilities that exist for community and individual action,such as establishing day care centres or encouraging the active participation of fathers in child care. *Health.* In addition to household food security and care,people have to be healthy to get the full nutritional benefits from food,otherwise some of the nutrients will be wasted.This means the nutrients are either passed through the digestive system unabsorbed,as in the case of someone suffering from diarrhoea,or consumed by parasites such as worms.This is why households must work to prevent communicable diseases and parasitic infections by practising good environmental and personal hygiene,by drinking safe water and keeping food safe and clean.If household members become sick,they should visit the health centre or seek advice from the village health worker.If not treated,infections that lead to poor utilization of available food and nutrients in the body can result in malnutrition.

Nutritional balance can be achieved through ensuring a variety of food and biological diversity

Variety of food and biological diversity. Planting small amounts of a wide variety of vegetables is a sound environmental strategy,as it conserves biological diversity and contributes to food security and nutrition.Women farmers are particularly interested in maintaining crop diversity,since they are often the ones who use these genetic resources to develop new varieties according to their families' changing needs and preferences.Growing a diversity of crops reduces the risk of a total crop failure as a result of disease.It also helps households vary their diets,prepare a broader range of dishes and enhance the nutritional quality of their meals.

During the rainy season,households have access to several traditional leafy vegetables that grow as either uncultivated or semi-cultivated vegetables on the farm,but these varieties may not be readily available during the dry season.Their inclusion,therefore,in the home garden cropping pattern,particularly during the dry season,supplements the dry-season vegetable supply and safeguards against crop failure,since traditional crops are well adapted to the local environment.To ensure the availability of seeds from such vegetables during the dry season,farmers must take care to collect their seeds at the end of the rains,store them properly and plant them in the home garden under irrigation.

IMPROVING THE CONTRIBUTION OF THE HOME GARDEN TO DAILY HOUSEHOLD FOOD NEEDS

OBJECTIVE

By the end of this session field workers will be able to identify the home garden technology options for improving the productivity of local home gardens and as a result the households' nutrition.

OVERVIEW

A well-developed home garden has the potential (when access to land and water is not a major limitation) to supply most of the non-staple foods and some of the staple foods that a family needs every day of the year.These gardens produce a wide variety of food crops that supply the households with vegetables and fruits,roots and tubers,legumes,spices,animals and,sometimes,fish.

This session covers the technology options and management advice for improving year-round access of households to a variety of nutritious foods.The problems affecting the food system of a given local community reviewed in Session 4,provide the starting point for discussions during this session.Based on their knowledge of the local communities and available technologies,the participants will look at each potential area of intervention and identify and review the various technology options that appear feasible within the local context. In addition to the Information Sheets studied in the previous sessions,participants should study those on soil management and fertility,water and crop management,and weed and pest management.

The results of the training needs assessment discussed in the Introduction to this manual should help trainers determine the amount of time needed for this session and whether to use Option 1 or Option 2 (explained in the following section).If the participants have sound horticultural backgrounds,Option 1 may be followed; if their horticultural knowledge needs strengthening,one to three additional days may be needed.The trainer with the agricultural background can determine the additional time required for this session.

ACTIVITIES

The trainers should have distributed Information Sheets 1,2 and 6 through 12,and all the Home Garden Technology Leaflets on the first day of the training course and asked the field workers to study a specified set (two to three leaflets per evening) each evening.Based on the training needs assessment results,the trainers then select one of the following two options.

*Option 1.*The participants divide into five groups.Each group studies one of the following sets of Information Sheets and the corresponding Home Garden Technology Leaflets,and prepares to answer questions about them.The trainers ascertain that all the participants are familiar with the contents of each leaflet.

*Option 2.*This option is the same as Option 1,with the inclusion of some field sessions on aspects of practical horticulture (if the training needs assessment results show that the participants need more practical background information).

IMPROVING AND MAINTAINING SOIL FERTILITY ARE CRITICAL TO IMPROVED CROP PRODUCTION IN THE HOME GARDEN

The improvement and maintenance of soil fertility are major challenges facing most home gardeners in Africa.They must,therefore,have access to the appropriate information and technical advice on:

- locally available organic matter that decomposes within a few weeks to make *compost*;
- *crop-rotation* and *intercropping practices* that minimize the depletion of plant nutrients in the soil;
- types of crops that can be planted as *green manure;*
- integration of crop and livestock production in a manner that creates *animal manure* best suited to the types of crops grown by the local community;
- *chemical fertilizers* (where affordable) that are essential for complementing compost and animal manure.

Home garden managers must have access to information on the fertilizer types best suited to specific home garden crops and the appropriate times for fertilizer application.

The participants should help farmers choose among the different technology and home garden management options.Where appropriate,participants can demonstrate soil management,soil fertility improvements and other relevant technologies,either on demonstration plots or,preferably,on the farmers' own land.The field workers should exploit all available means of disseminating information,including radio,television,newspapers,songs and local theatre.

SELECTION OF THE TECHNOLOGIES AND PLANT VARIETIES

In addition to soil management and the improvement of soil fertility,a wide range of technologies exists for improving home garden production and productivity.The area's agro-ecological and climatic characteristics and its altitude largely determine the types of crops to plant and the seasons during which those crops grow well.Technology options,therefore,should be directly related to the altitude and agro-ecological and climatic features of a given area.Farmers,therefore,require information on:

- the most appropriate water management and water-harvesting techniques;
- the crop varieties best suited to their agro-ecological zones and the seasons during which those crops grow well;
- low-cost and sustainable pest and weed management techniques.

Water management and water harvesting. The annual rainfall pattern and access to reliable watering facilities often determine the physical location of a home garden and the types of beds a farmer may construct.Farmers must have access to appropriate information in order to know whether to construct sunken,flat or raised beds or mounds,depending upon which crops and which times of the year.

Regions with steady rainfall throughout the year have greater potential for producing a steady,year-round supply of fruits and vegetables.On the other hand,communities in areas with distinct wet and dry seasons must have a variety of strategies for ensuring an adequate,year-round supply of fruits and vegetables.Such strategies can include developing two household garden plots,one around the homestead and exploitable mainly during the rains (i.e.rainfed),the other one away from home in a wet lowland or swampy place.For the household garden plot to produce vegetables and other crops throughout the dry season,the use of unconventional water sources,such as household wastewater (from washing dishes or bathing) might be an option.

In addition,communities in areas with inadequate access to dry-season irrigation water can secure a communal water point,by constructing small dams or hand-dug wells,paid for by the pooling of community resources.Next to this water source,participating households can plant a "community garden" in which each household has its individual plot but shares the water and fence with the other participating community members.Assisted by the field workers,community members can jointly decide on the community garden's planting patterns and times for specific food crops.Seed and other gardening inputs can be purchased with help from a community fund or with savings accumulated through group savings schemes.Such community gardens provide an excellent contact point for the delivery of technical information by the agricultural,nutritional or health extension workers in the area.

*Crops and crop varieties must be suited to the environment.*While the integration of traditional and local vegetables into the home garden cropping pattern should be encouraged,the low yields of such crops inhibit productivity increases.However,home garden managers can be taught to maintain balance between the traditional and the improved,or introduced,varieties.

Encouraging the cultivation of a broad variety of food crops is a greater challenge in the semi-arid and arid regions and under low-temperature conditions (e.g.in highland areas) than it is in the humid areas.Efforts to extend the production of fruits and vegetables into the drier and colder periods of the year have to be based on the selection of hardier crop species and varieties,together with the application of appropriate cultural practices such as:

- raising early seedlings for transplanting;
- using windbreaks;
- adding mulch and organic matter to improve the soil's water-holding capacity;
- protecting crops from low temperatures with the use of plastic sheeting or straw;
- using all available water supplies to keep crops growing during dry periods.

*Low-cost pest management.*The control of plant pests and insects,particularly during dry-season home gardening,is another challenge of rural households.Farmers must have access to information on developing good home garden hygiene.In addition,they must be aware of environmentally friendly methods of controlling pests,for example:

- intercropping food crops with chilli,garlic,basil and other plants that repel certain pests;
- using natural pesticides (e.g.wood ash,tobacco,neem,papaya leaves) or easily made preparations such as soap or vegetable oil emulsions or starchy suspensions;
- removing and burning diseased plants;
- intercropping and rotating crops of different family types to check the spread of diseases and pests specific to a particular crop type.

A carefully selected planting pattern

When planning home garden improvements for a continuous supply of fruits and vegetables,it is important to take into account untapped resources and potentials as well as household food needs and preferences.

Factors such as the length of a crop's growing period (from planting to maturity) and the length of time during which a household may harvest edible fruits or vegetables from a mature plant before it must be replaced should be considered when planning a home garden planting pattern.In addition to taking into account basic crop-rotation requirements and desired intercropping aspects,home garden managers must stagger the planting of different types of crop.

To reduce gaps in the year-round supply of fresh fruits to the household,for example,the home garden manager must select and plant fruit-tree types with fruits that mature at different times during the year.These can be mixed with fruit-trees that bear fruit throughout the year (e.g.bananas).

Leafy vegetables and legumes must be planted in a manner that provides the household with access to some varieties of green leafy vegetables and fresh or dry legumes throughout the year.Vegetables that grow well all year round(e.g.some varieties of leafy vegetables) can form the core of the home garden,and seasonal varieties can be planted when appropriate.

Information on the types of fruits,vegetables and legumes that give the best yields at specific times of the year is crucial,because it enables the home garden manager to organize planting patterns for optimum yields and nutritional benefits.Home garden managers should also be sensitive to the planting patterns of other local farmers and,thus,can avoid flooding the market with one or two food commodities,should their household produce surplus for sale.

Planting a mixture of annual and perennial crops (e.g.cassava and fruit-trees) and adopting multiple and multistorey cropping techniques increase the year-round availability of food to the household.Sweet potato and cowpea can be useful cover crops.Small numbers of different crops and fruit-trees,and fresh vegetables in particular,planted and harvested consecutively,provide a constant supply of home garden foods. If adequate quantities and varieties of home garden foods are to be produced,particularly in the wet lowlands where year-round food production may be possible,intensive crop production techniques must be considered.The use of better cultural practices (i.e.soil fertility and water management,optimum planting times and plant spacing,proper crop rotations,weed control and integrated pest management) can increase the productivity of individual crops.As more intensive crop production practices are adopted,better post-harvest handling techniques (i.e.processing,preservation and storage) must be developed in order for farmers to gain maximum benefits from the increased garden output.

Finally,households should protect their home gardens from roaming animals,especially during the dry season.Using live fencing for this purpose will provide further benefits to the household (e.g.fuelwood,trellising and medicines).

The agricultural extension worker can help families and communities identify the crops that are suited to local conditions and that will supply the nutrients needed for the family to achieve optimum nutritional balance year round.Since women are usually responsible for home gardens,their full participation in planning and implementing improvements is essential.

PROMOTION OF HOME GARDENS FOR IMPROVED NUTRITION: SECOND HOME GARDEN VISIT

OVERVIEW

It is important to remember that the field worker's role is to facilitate learning and analysis by local households and communities.The emphasis,therefore,is on building,step by step,a process of discussion,analysis and planning in the household or community.This requires a number of follow-up visits and the establishment of a working relationship with local people on a continuing basis.Such a process requires teamwork,sensitivity and certain skills in facilitation.Gender roles and responsibilities,and the access by men and women to resources,must be included in the dialogue with households and communities,since these issues significantly affect the decisions that household members make about food and nutrition.

ACTIVITIES

Review of technical notes. Using the Technical Notes section in Session 7,the trainer and participants review the importance of teamwork,the facilitation and communication skills that field workers need in their work with households and communities,and the importance of considering gender issues in all agriculture-,home gardening- and nutrition-related activities at the household level.

Review and final preparation. Based on the outcome of the first home garden visit and the subsequent training sessions,the trainer facilitates a discussion that leads.

The trainer emphasizes the importance of treating the checklist as a guide for discussion rather than as a rigid questionnaire.

Preparation for household visits. The trainer should arrange to re-visit the same households that participated in the first home garden visit.For households other than single-headed ones,both the husband and the wife should be present.The field assessment will take an entire day.

Field workers continue to work in the same groups as during previous visits and classroom exercises.Each group selects a facilitator and a recorder prior to the visit.Each team will interview five to six households.The health worker or nutritionist in the group will be responsible for the nutritional assessment (measuring MUAC of children aged one to five years).

Second home garden visit. After introductions,the group facilitator explains the purpose of the visit and provides a brief summary of the results of the first visit.Each team and its respective households begin discussions.

The participants explore the home garden with the household members and ask household members to help them complete Parts E,F,G and H.

The team leader facilitates a session with the household where all members participate in drawing a map of the home garden (on a flip chart or on the ground),identifying its main physical features and the crops grown there.This map will serve as a reference for discussion in this session and in Session 8.

FIELD WORKERS NEED GOOD FACILITATION AND COMMUNICATION

Apart from having specific technical knowledge and skills,field workers who work directly with households and communities must also have excellent interpersonal and facilitation skills.A good facilitator is a good observer and an active listener,who shows interest in what people say.A good facilitator should generally:

- be respectful of local opinions;
- quickly identify individuals who dominate discussions and find gentle and tactful ways of soliciting the opinions of other community members,in group discussions or individually;
- be observant of group dynamics and how household and community members (e.g.husband and wife or men and women) interact;
- respond to the needs of households and adapt questions according to those needs.
- The field workers should generally be sensitive to cultural practices and traditions in their dealings with households.For example,where it is not acceptable for male field workers to meet with female members of a household or when the topic of discussion is considered

exclusively "female" (e.g.food preparation,child feeding,reproductive health or family planning),female field workers should facilitate the discussion or organize women-only focus group sessions.That is why it is important that each team include female members,preferably in equal proportion to the male members.

TEAMWORK AMONG FIELD WORKERS IS CRUCIAL FOR HELPING HOUSEHOLDS IMPROVE NUTRITION

Food and nutrition issues cut across several disciplines,and helping households solve problems of food insecurity and malnutrition,becomes the responsibility of all field workers working in a community.When solving these problems individually,agricultural extension workers,for example,normally look only at production,crop yields,prices and income,while nutritionists and health workers,on the other hand,consider how households acquire food and whether that food is adequate to meet people's nutritional needs.Creating an environment in which field staff from different sectors can share information and work together for a common goal ensures that rural households are able to draw upon available technical services and inputs in a manner that addresses the multifaceted nature of food insecurity and poor nutrition.

Considering gender roles during dialogue with community members is of key importance

Gender roles (i.e.the roles and responsibilities held by both men and women) have to be taken into account during dialogues aimed at improving household nutrition.This can be done by looking at questions such as:

- Who decides what to grow,where to grow it (home garden or family field),when to grow it and what to do with the food produced?
- Who is responsible for the day-to-day decisions about the types of food to cook,how much food to cook and how much to serve to the different members of the household?
- How do different members of the household spend their days (daily activity patterns) and what are the seasonal variations in these activity patterns?
- What are the specific needs of men and women in home garden and nutrition improvement and how can those needs be met?

A technique such as asking men and women to keep separate daily logs of their activities will provide a clear indication of who does what and when,and will indicate the competing time demands on women for making home garden improvements and providing adequate care for young children.

It may be necessary for the participants to meet separately with women.This will allow the women to express their ideas and opinions freely,especially in places where it is culturally unacceptable for women to speak in the presence of men or at public gatherings.The active participation of women in assessing the food and nutrition situation is important,both as a means of empowering women and of ensuring that home garden actions and improvements will be designed and planned in a way that meets the women's needs without burdening them with extra tasks.

IMPROVING NUTRITION THROUGH HOME GARDENING: DEVELOPING A PLAN OF ACTION

Fig. Husband,wife and children discussing home garden improvements

OVERVIEW

Individual consultations with households are valuable but cannot give families all the knowledge and skills they need to make their own decisions and then act on them.The community,as a group or in several groups,must take actions that make people aware of the importance and urgency of improved food production and better nutrition,and then be assisted in implementing those actions.Information and practical training in home garden technologies and nutrition are needed to achieve these goals.This is where agricultural extension workers and other field workers can provide valuable support.

Proper action requires good planning,and good planning can be done by using a simple project planning method.In this session,the basic terminology and principles of project planning and formulation are discussed,illustrated and applied.

ACTIVITIES

*Presentation of method for project planning.*The trainer uses the Technical Notes in this session to introduce the elements of a project.He or she emphasizes that the project method is based on simple reasoning and is particularly valuable because it guides people in preparing for needs and any problems that might arise after work on the project has begun.

*Preparing a project.*The trainer displays the first three stages of Project Outline 1: Training in home garden development which are the rationale,goal and objective of the project.The trainer asks the field workers to work in small groups to finish writing the plan of Project Outline 1 for their given area and to think about how an intervention or project could be implemented.Field workers can draw upon the information acquired during their dialogues with community members.Their work should include discussion and preparation of the:

- strategy;
- expected outputs;
- activities and inputs;
- timetable;
- indicators for project monitoring and evaluation.

The participants can propose the project elements in any order,but they should organize and present them in a logical manner.Depending on the exposure and project formulation experience of the field workers,the assistance of the trainer may be required.

*Feedback.*After the participants have completed their project outlines,the trainer displays the completed Project Outline 1 and invites the field workers to discuss the differences between it and their own project outlines.

Note: Project Outline 1 is only an example.The participants' own project outlines will probably be more sensitive to local needs and conditions.

*Small group discussion of indicators.*After going through the elements of the project,the trainer emphasizes the importance of selecting indicators.Indicators will help the community and participants determine if the selected home garden and nutrition improvement activities are producing the expected results.The trainer then divides the participants into small groups.

COMMUNITY ACTION MUST RAISE AWARENESS AS WELL AS PROVIDE PRACTICAL TRAINING

In previous sessions,the participants learned to recognize the importance of household food security (i.e.having access to enough nutritious foods year round) and the home garden's potential contribution to food security.They applied their knowledge both to case studies and to individual real-life situations.

The next thing to consider is how the entire community can improve its household food supply.An initial step towards developing the different components of a community plan of action is to list the solutions that community members consider feasible.Households may be able to implement some of these solutions immediately,while other solutions may require them first to mobilize resources and acquire practical skills. There also may be a set of problems that community members do not necessarily perceive as problems because they lack the knowledge (e.g.the causal relationship between parasitic diseases and malnutrition).Participants should identify these types of knowledge gaps.

The strategy for action to be adopted must therefore involve:

- raising community and household members' awareness of the importance and urgency of improving a home garden's production for better nutrition and extra income;
- encouraging and supporting them,through practical skills training,in taking immediate action using existing resources.

With the list of activities established,community members can discuss them and then decide jointly how,by whom,when and where those activities should be undertaken,and what inputs and technical assistance may be required to implement them.Knowledge of the different elements and stages of planning enables field workers to facilitate this step-by-step process.

Action to promote improved nutrition through home gardening requires careful planning

Any project or programme involving a number of households will need careful planning.One of the most effective ways of planning an action programme is to work systematically through the elements of project formulation:

- Rationale;
- Goal;
- Objective;
- strategy;
- outputs and targets;
- activities;
- inputs;
- timetable and work plan;
- monitoring; evaluation.

Planning a project is like planning a journey.Before you embark,you need to decide:

- why you are going (*rationale* and *goal*);
- where you want to go (*objective*);
- how you propose to get there (*strategy*);
- what the immediate results of your trip will be (*output*);
- when you expect to arrive (*target*);
- what preparations you need to make and how you will travel (*activities*);
- what you will need,e.g.documents,vehicles,petrol (*inputs*);
- how long it will take (*timetable*);
- what progress is being made along the way (*monitoring*);
- has the objective been achieved (*evaluation*).

A methodical approach makes it possible to:

- determine how realistic the project is and whether it can be implemented with locally available resources and expertise;

- anticipate problems;
- check on the progress of the work (monitoring) and on whether the project has achieved its aims (evaluation).

ELEMENTS OF A PROJECT

Rationale. A rationale is the reasoning behind a plan.In the case of home garden development,the lack of knowledge and skills among the household members means that some form of education is important.The rationale should also clarify who is to benefit from the plan.Women are often the managers of home gardens and the household food supply.Therefore,women and their husbands may be the direct target group and beneficiaries of home garden improvement projects,while children will be the main indirect beneficiaries because they will benefit from an improved diet.

Goal. A goal is an ultimate aim.In this training course,the goal is improved food production and nutrition through better use of the home garden.

Objective. An objective is what is hoped will happen as a result of the project activities (e.g.households will improve their farming methods in specific ways,they will produce more foods and a wider range of foods,and families will have better diets).More specific than a goal,an objective should be worded in such a way as to make it possible to measure (in quantitative terms) the progress towards it.

Strategy. A strategy is a general procedure proposed for achieving an objective (e.g.,a community group will decide whether it will implement improvements as a group or as individual households,which activities/services can be organized jointly and which individually,and how group members can assist one another in technology transfer and information sharing).The strategy should take into account all local constraints and conditions,including traditional and cultural factors as well as resource and climatic constraints.

Outputs. An output is an actual result or product of project activities (e.g.20 households trained in soil and water management,one demonstration home garden developed,20 water- harvesting facilities established).If there are many activities to achieve one objective,there will be many outputs for that one objective.In any detailed plan,the description of an output should include the quantities involved and a date by which the output must be completed.

Activities. Activities are what is done during the project to produce the outputs (e.g.training,demonstrating,selecting field workers,producing charts).In a detailed plan,it is important that the description of activities make it clear who is responsible for each activity.

Inputs. Inputs are all the things (e.g.labour,equipment,money) necessary to make the activities possible.In a detailed plan,inputs should bequantifie (e.g.30notebooks, 2 trips,10 work days per person) and their costs estimated.

*Timetable,or work plan with a time frame.*The timetable sets out the time needed to complete each of the project activities and makes it possible to measure progress during implementation. It can be prepared as a detailed work plan showing who does what,when,over what period and with what results.

*Monitoring.*This is an essential component of each project.It involves recording specific types of information (i.e.indicators) on a regular basis to enable group members to assess the progress they are making towards reaching their objective.It can also be an important management tool.It helps determine if the objectives are realistic or if they need to be revised,and helps identify and anticipate problems so that group members can take steps to avoid or solve them.

*Evaluation.*An evaluation is done at the end,or when a project is completed.It measures whether the project has achieved its stated aims or objectives.Well-defined objectives are important to allow measurement of success. Example of a project outline.The project elements described above are illustrated in Project Outline 1,an example of a project outline on training in home garden development. This is only an example,and the field workers can prepare different ones based on the outcome of the follow-up field visit.

RATIONALE

Home gardening can be greatly improved by enhancing the use of the garden for improved household food security and nutrition.Many people in rural villages lack the experience,knowledge and skills to make improvements. Sometimes they also lack appropriate planting materials. Knowledge and skills can be transferred by identifying successful home garden managers who are prepared to work with less-successful or less-knowledgeable households, supported by agricultural extension staff. The need for planting materials can be met by setting up nurseries in individual home gardens,which can be managed by the respective managers on a commercial basis.Visible success is more persuasive than a thousand lectures.

GOAL

The goal is to improve food production,supply and utilization and,ultimately,the nutritional well-being of the community through better use of its home gardens.

OBJECTIVE

The objective is to enable 200 households to prepare and put into action their own plans for improving year-round availability of a variety of nutritious foods by (one or several of these,depending on the local conditions):

- diversifying crops;
- improving water and soil management;
- intensifying land use;

- improving crop protection;
- improving food processing,preservation and storage;
- improving the use of garden produce during the preparation of family meals.

STRATEGY

A field worker will help families analyse the main problems that exist in their home gardens and the options and opportunities they have for making improvements (e.g.available land and sufficient labour).The field worker will also help identify those families who have developed their home gardens well and who apply good practices. Home garden managers (e.g.husband or wife,or both) will be invited to a series of training sessions involving demonstrations of good home garden practices and the introduction of new techniques.With the help of the field worker,these home garden managers will then develop plans for improving their own home gardens.In addition,practical demonstrations of food processing and preparation will show them how to make the best use of their produce for the nutritional well-being of their families.

OUTPUTS AND TARGETS

Note: For each of the following outputs and targets,a date of completion should be set.The expected outputs include:

- Main home garden problems identified by home garden owners or managers (with the help of the field worker);
- well-developed local home gardens identified;
- training needs of home garden managers identified and assessed;
- a training plan and materials prepared;
- home garden managers (e.g.10 groups of 20) trained in improving home garden management as well as in other aspects identified as part of the assessment;
- individual home garden improvement plans developed and a list of inputs and materials prepared,including the cost of those inputs,by the home garden managers;
- indicators selected for monitoring progress during the implementation of activities.

ACTIVITIES

The field worker

When planning activities for each output,the field worker:

- assists home garden managers (i.e.during the household consultation) in appraising their home gardens and identifying the potential for increased food production and the problems that need to be addressed;

- identifies two or three home gardens that show good soil and water management,diversified and intensified cropping systems and good individual crop productivity;
- assists home garden managers in identifying their need for knowledge and skills training;
- prepares the appropriate training sessions and training materials,ensuring that the location and timing of the training will suit the needs of the home garden managers;
- invites home garden managers to participate in the training course;
- conducts five 2-hour training sessions (including one walking tour of a good example of a well-developed home garden,using that garden's owner as a guide,one demonstration session on easy new techniques, two follow-up discussion sessions and one session for finalizing individual home garden plans);
- arranges supplies of seeds and seedlings,as necessary;
- visits community members' home gardens to facilitate and guide them in implementing their plans.

Community members

- When planning activities for each output,the community members:
- identify the problems and potentials of individual home gardens;
- develop plans for the improvement of those gardens;
- participate in training on relevant technology options for home garden improvement;
- implement the selected actions;
- monitor progress;
- participate in evaluating the results.

In the above examples,the activities are classified by the people or groups of people who will be doing them,in order to show clearly each group's different responsibilities.

Normally under each output there would be only one list of activities and an indication of who will be responsible for each activity and the time it should take to be carried out.Based on this information,a work plan is prepared.

INPUTS

The following inputs are needed:

- field worker (25 days of work);
- seeds and seedlings from a local nursery to be purchased by home garden managers;
- technical support
- (one or two advisers for training sessions) from the given country's Ministry of Agriculture.

TIMETABLE

The timetable usually indicates the starting time and duration of an activity:

- information collection: ten days;
- planning of training: five days;
- training: five days;
- follow-up visits: five days.

Note: After listing the inputs,the next questions are,"Where will the inputs come from,and what will be their cost?" Such information is useful in identifying low-cost options that can be implemented immediately.Options that are costly (e.g.building a water well) may require joint action by the community.In this case,the field worker must assist the community in preparing a proposal,including a budget,for submission to a community development fund (operated by the government or an NGO) that provides matching grants for community projects.Matching grants provide funds or inputs (e.g.materials,equipment,technical assistance) that must be "matched" by a community contribution in kind,for example,the labour needed to dig a well and materials that the community can supply from its own resources (e.g.cement and sand for making bricks).

BUDGET

After listing those inputs that must be costed,the next question is,"Which inputs can be met from community resources and which require outside support?" Such information is useful in preparing a budget.

PREPARING PLANS OF ACTION WITH THE COMMUNITY

Fig. Preparing plans of action with the community

OVERVIEW

The principles studied in the previous session are now applied in the designing of real-life plans of action for local needs and conditions.Some of the interventions can be planned and implemented by individual households,while others can be done collectively.The option chosen largely depends on what the community members decide to do and how they organize themselves.

ACTIVITIES

Selection of priority issues. Depending on the local potentials and limitations,community members are asked to choose the solutions that they want to implement first and list them in the order of importance.Interventions given equal priority can be placed together on the list.

Preparing the project and work plan. The field workers will then facilitate community discussions and work with the community step by step in the various elements of the plan of action.To facilitate dialogue in preparing the planning of each of the selected interventions,field workers work through the following list of questions with a community group:

- Why do you want to implement the intervention (*rationale* and *goal*)?
- What do you want to achieve (*objective*)?
- How will you implement the intervention (*strategy*)?
- What immediate results of the various activities do you foresee (*output*)?
- What must be done and by whom (*activities*)?
- What materials (e.g.planting materials,technical advice,equipment) will be needed in order for the foreseen actions to take place (*inputs*)?
- Which items or services must be purchased,and at what cost (*budget*)?
- Who will pay for these items or services (*source of inputs/finance*)?
- When will each of the foreseen activities begin and how long will it take for each of them to be completed (*timetable or work plan*)?
- What problems do you foresee during implementation and how can these problems be avoided or minimized (*risk factors*)?
- How will the community check to see if the intervention is progressing in the right direction (*monitoring*)?
- How will the community determine if the intervention has achieved its intended aim (*evaluation*)?

Further refinement of the plan.During the process of discussing and responding to the above questions,the community finally decides whether or not a particular solution is realistic,particularly after establishing the cost of the inputs.

This enables community members to decide if they can afford the option in the short term or if they will have to wait until they have raised enough resources from within or outside the community.

Follow-up to the development of the community plan of action.After finalizing the community plan of action,each group or person involved or responsible for implementing a particular activity can begin to take action,as indicated in the work plan.The community members need to set up a mechanism for ensuring that activities that involve different sectors and responsibilities (e.g.a cooking demonstration to which the field workers - health workers,home economists and nutritionists in particular - and participating households will be invited,and for which cooking utensils and food items must be on hand) are properly coordinated by the person appointed for this task by the community.The field workers should provide technical support and assist the communities in monitoring progress,using indicators selected by the community to do this.

Evaluation of the field visit and evaluation of the training.At the end of the field visit,field workers evaluate the day's activities and discuss the follow-up action to be taken upon returning to their local communities.

THE MOST EFFECTIVE STRATEGY IS TO WORK WITH THE PEOPLE AND RESOURCES AVAILABLE

In order to be successful,community projects must be planned together with the households concerned.They also must work within the limits of available resources.Some examples of local resources commonly available are:

- well-developed home gardens in the area;
- local skills in,knowledge of and good practices for home gardening,food processing and preparation;
- natural resources,such as land,soils,water and forests;
- local nurseries;
- local research and extension service stations of the given country's Ministry of Agriculture;
- community groups and associations;
- the local government;
- NGOs;
- local colleges or universities;
- information collected during the home garden visits.

6

Elements of Gardening and Design

GARDEN DESIGN

Fig. The White Garden at Sissinghurst Castle Garden,designed by Vita Sackville-West

Garden design is the art and process of designing and creating plans for layout and planting of gardens and landscapes.Garden design may be done by the garden owner themselves,or by professionals of varying levels of experience and expertise.

Most professional garden designers have some training in horticulture and the principles of design,and some are also landscape architects,a more formal level of training that usually requires an advanced degree and often a state license.Amateur gardeners may also attain a high level of experience from extensive hours working in their own gardens,through casual study,serious study in Master Gardener Programs,or by joining gardening clubs.

ELEMENTS OF GARDEN DESIGN

Whether a garden is designed by a professional or an amateur,certain principles form the basis of effective garden design,resulting in the creation of gardens to meet the needs,goals and desires of the users or owners of the gardens. Elements of garden design include the layout of hard landscape,such as paths,walls,water features,sitting areas and decking; as well as the plants themselves,with consideration for their horticultural requirements,their season-to-season appearance,lifespan,growth habit,size,speed of growth,and combinations with other plants and landscape features.Consideration is also given to the maintenance needs of the garden,including the time or funds available for regular maintenance,which can affect the choice of plants in terms of speed of growth,spreading or self-seeding of the plants,whether annual or perennial,and bloom-time,and many other characteristics.

Important considerations in garden design include how the garden will be used,the desired stylistic genre (formal or informal,modern or traditional etc.),and the way the garden space will connect to the home or other structures in the surrounding areas.All of these considerations are subject to the limitations of the prescribed budget.

LOCATION

A garden's location can have a substantial influence on its design.Topographical landscape features such as steep slopes,vistas,hills and outcrops etc.may suggest or determine aspects of design such as layout,and can be used and augmented in order to create a particular impression.The soils of the site will affect what types of plant may be grown,as will the garden's climate zone and various microclimates.The locational context of the garden can also influence its design; for example an urban setting may require a different design style to a rural one.Similarly,a windy coastal location may necessitate a different treatment compared to a sheltered inland site.

Soil

The quality of a garden's soil can have a significant influence on a garden's design and its subsequent success.Soil influences the availability of water and nutrients,the activity of soil micro-organisms,and temperature within the root zone,and thus may have a determining effect on the types of plants which will grow successfully in the garden.However soils may be replaced or improved in order to make them more suitable.

Traditionally,garden soil is improved by amendment,the process of adding beneficial materials to the native subsoil and particularly the topsoil.The added materials,which may consist of compost,peat,sand,mineral dust,or manure,among others,are mixed with the soil to the preferred depth.The amount and type of amendment may depend on many factors,including the amount of existing soil humus,the soil structure (clay,silt,sand,loam etc.),the soil acidity/ alkalinity,and the choice of plants to be grown.One source states that,"conditioning the soil thoroughly before planting enables the plants to establish themselves quickly and so play their part in the design." However,not all gardens are,or should be,amended in this manner,since many plants prefer an impoverished soil.In this case,poor soil is better than a rich soil that has been artificially enriched.

Boundaries

The design of a garden can be affected by the nature of its boundaries,both external and internal,and in turn the design can influence the boundaries,including via creation of new ones.Planting can be used to modify an existing boundary line by softening or widening it.Introducing internal boundaries can help divide or break up a garden into smaller areas.

The main types of boundary within a garden are hedges,walls and fences.A hedge may be evergreen or deciduous,formal or informal,short or tall,depending on the style of the garden and purpose of the boundary.A wall has a strong foundation beneath it at all points,and is usually - but not always - built from brick,stone or concrete blocks.A fence differs from a wall in that it is anchored only at intervals,and is usually constructed using wood or metal (such as iron or wire mesh).

Boundaries may be constructed for several reasons: to keep out livestock or intruders,to provide privacy,to create shelter from strong winds and provide microclimates,to screen unattractive structures or views,and to create an element of surprise.

Surfaces

Fig. Naturalistic planting design

In temperate western gardens,a smooth expanse of lawn is often considered essential to a garden.However garden designers may use other surfaces,for example those “made up of loose gravel,small pebbles,or wood chips” in order to create a different appearance and feel.Designers may also utilise the contrast in texture and color between different surfaces in order to create an overall pattern in the design.

Surfaces for paths and access points are chosen for practical as well as aesthetic reasons.Issues such as safety,maintenance and durability may need to be considered by the designer.Gardens designed for public access need to cope with heavier foot traffic and hence may utilise surfaces - such as resin-bonded gravel - that are rarely used in private gardens.

Planting design

Planting design requires design talent and aesthetic judgement combined with a good level of horticultural,ecological and cultural knowledge.It includes two major traditions: formal rectilinear planting design (Persia and Europe); and formal asymmetrical (Asia) and naturalistic planting design.

History

Persian gardens are credited with originating aesthetic and diverse planting design.A correct Persian garden will be divided into four sectors with water being very important for both irrigation and aesthetics.The four sectors symbolize the Zoroastrian elements of sky,earth,water and plants.Planting in ancient and Medieval European gardens was often a mix of herbs for medicinal use,vegetables for consumption,and flowers for decoration.Purely aesthetic planting layouts developed after the Medieval period inRenaissance gardens,as are shown in late-renaissance paintings and plans.The designs of the Italian Renaissance garden were geometrical and plants were used to form spaces and patterns.The gardens of the French Renaissance and Baroque Garden à la française era continued the ‘formal garden’ planting aesthetic.

In Asia the asymmetrical traditions of planting design in Chinese gardens and Japanese gardens originated in the Jin Dynasty (265–420) of China.The gardens’ plantings have a controlled but naturalistic aesthetic.In Europe the arrangement of plants in informal groups developed as part of the English Landscape Garden style,and subsequently theFrench landscape garden,and was strongly influenced by the picturesque art movement.

Application

A planting plan gives specific instructions,often for a contractor about how the soil is to be prepared,what species are to be planted,what size and spacing is to be used and what maintenance operations are to be carried out under the contract.Owners of private gardens may also use planting plans,not for contractual purposes,as an aid to thinking about a design and as a record of what has been planted.A planting strategy is a long term strategy for the design,establishment and management of different types of vegetation in a landscape or garden.

Fig. Garden chairs and table

Planting can be established by directly employed gardeners and horticulturalists or it can be established by a landscape contractor (also known as a landscape gardener).Landscape contractors work to drawings and specifications prepared by garden designers or landscape architects.

Garden furniture

Garden furniture may range from a *patio set* consisting of a table,four or six chairs and a parasol,through benches,swings,various lighting,to stunning artifacts in brutal concrete or weathered oak.Patio heaters,that run on bottled butane or propane,are often used to enable people to sit outside at night or in cold weather.A picnic table,is used for the purpose of eating a meal outdoors such as in a garden.

The materials used to manufacture modern patio furniture include stones,metals,vinyl,plastics,resins,glass,and treated woods.

Sunlight

While sunlight is not always easily controlled by the gardener,it is an important element of garden design.The amount of available light is a critical factor in determining what plants may be grown.Sunlight will,therefore,have a substantial influence on the character of the garden.For example,a rose garden is generally not successful in full shade,while a garden of hostas may not thrive in hot sun.As another example,a vegetable garden may need to be placed in a sunny location,and if that location is not ideal for the overall garden design goals,the designer may need to change other aspects of the garden. In some cases,the amount of available sunlight can be influenced by the gardener.The location of trees,other shade plants,garden structures,or,when designing an entire property,even buildings,might be selected or changed based on their influence in increasing or reducing the amount of sunlight provided to various areas of the property.

In other cases,the amount of sunlight is not under the gardener's control.Nearby buildings,plants on other properties,or simply the climate of the local area,may limit the available sunlight.Or,substantial changes in the light conditions of the garden may not be within the gardener's means.In this case,it is important to plan a garden that is compatible with the existing light conditions.

Lighting

Garden lighting can be an important aspect of garden design.In most cases,various types of lighting techniques may be classified and defined by heights: safety lighting,uplighting,and downlighting.Safety lighting is the most practical application.However,it is more important to determine the type of lamps and fittings needed to create the desired effects.Light regulates three major plant processes: photosynthesis,phototropism,and photoperiodism.

Photosynthesis provides the energy required to produce the energy source of plants. Phototropism is the effect of light on plant growth that causes the plant to grow toward or away from the light.Photoperiodism is a plant's response or capacity to respond to photoperiod,a recurring cycle of light and dark periods of constant length.

TYPES OF GARDENS

Fig. Moorish Generalife courtyard fountain

Islamic gardens

Garden design,and the Islamic garden tradition,began with creating the Paradise garden in Ancient Persia,in Western Asia.It evolved over the centuries,and in the different cultures Islamic dynasties came to rule in Central— South Asia,the Near East,North Africa,and theIberian Peninsula.

Examples

Some styles and examples include:

1. Persian gardens
- Eram Garden
- Fin Garden
2. Mughal gardens
- Nishat Bagh
- Shalimar Gardens (Lahore)
- Yadavindra Gardens (Pinjore)
- Charbagh
- Taj Mahal
- Tomb of Humayun gardens
- Bagh (garden)
- Bagh-e Babur
- Shalimar Bagh (Srinagar)
- Al-Andalus—Moorish architecture and gardens
- Alcázar of Seville
- Alhambra
- Generalife

Mediterranean gardens

Garden design history and precedents from the Mediterranean region include:

- Ancient Greek and Hellenistic gardens
- Ancient Roman gardens
- Peristyle gardens —*evolved into Cloister gardens.*
- House of the Vettii — *Pompeii.*
- Horti Sallustiani
- Byzantine gardens
- Spanish gardens
- Andalusian Patio

Renaissance and Formal gardens

Main article: Renaissance gardens

A formal garden in the Persian garden and European garden design traditions is rectilinear and axial in design.The equally formal garden,without axial symmetry (asymmetrical) or other geometries,is the garden design tradition of Chinese gardens and Japanese gardens.The Zen garden of rocks,moss and raked gravel is an example.The Western model is an ordered garden laid out in carefully planned geometric and often symmetrical lines.Lawns and hedges in a formal garden need to be kept neatly clipped for maximum effect.Trees,shrubs,subshrubs and otherfoliage are carefully arranged,shaped and continually maintained.A French garden or Garden à la française,is a specific kind of formal garden,laid out in the manner of André Le Nôtre; it is centered on the façade of a building,with radiating avenues and paths of gravel,lawns,parterresand pools (*bassins*) of reflective water enclosed in geometric shapes by stone coping,with fountains and sculpture.

The Garden à la française style has origins in fifteenth-century Italian Renaissance gardens,such as the Villa d'Este,Boboli Gardens,and Villa Lante in Italy.The style was brought to France and expressed in the gardens of the French Renaissance.Some of the earliest formal parterres of clipped evergreens were those laid out at Anet by Claude Mollet,the founder of a dynasty of nurserymen-designers that lasted deep into the 18th century.The Gardens of Versailles are an ultimate example of Garden à la française,composed of many different distinct gardens,and designed by André Le Nôtre.

English Renaissance gardens in a rectilinear formal design were a feature of the stately homes.The introduction of the parterre was at Wilton House in the 1630s.In the early eighteenth century,the publication of Dezallier d'Argenville,*La théorie et la pratique du jardinage* (1709) was translated into English and German,and was the central document for the later formal gardens of Continental Europe.

Traditional formal Spanish garden design evolved with Persian garden and European Renaissance garden influences.The internationally renowned Alhambra and Generalife inGranada,built in the Moorish Al-Andalus era,have influenced design for centuries.The Ibero-American Exposition of 1929 World's Fair in Seville,Spain was located in the celebrated Maria Luisa Park (*Parque de Maria Luisa*) designed by Jean-Claude Nicolas Forestier.

Formal gardening in the Italian and French manners was reintroduced at the turn of the twentieth century.Beatrix Farrand's formal Italian garden areas at Dumbarton Oaks inWashington,D.C.,and Achille Duchêne's restored French water parterre at Blenheim Palace in England are examples of the modern formal garden.The Conservatory Garden in Central Park of New York City features a formal garden,as do many other parks and estates such as Filoli in California.

The simplest formal garden would be a box-trimmed hedge lining or enclosing a carefully laid out flowerbed or garden bed of simple geometric shape,such as a knot garden.The more developed and elaborate formal gardens contain statuary and fountains.

Fig. Formal garden laid out at the Abbaye de Valloires,Picardy,by Gilles Clément,1987

Features in a formal garden may include:

- Terrace
- Topiary
- Statuary
- Hedge
- Bosquet
- Parterre
- Sylvan theater
- Pergola
- Pavilion
- Landscaping

English Landscape and Naturalistic gardens

The English Landscape Garden style practically swept away the geometries of earlier English and European Renaissance formal gardens.William Kent and Lancelot "Capability" Brown were leading proponents,among many other designers.The naturalistic English Garden style (French: *Jardin anglais*,Italian: *Giardino all'inglese*,German: *Englischer Landschaftsgarten*) of the 1730s and on transformed private and civic garden design across Europe.The French Landscape Garden subsequently continued the style's development on the Continent.

Cottage gardens

A cottage garden uses an informal design,traditional materials,dense plantings,and a mixture of ornamental and edible plants.Cottage gardens go back many centuries,but their popularity grew in 1870s England in response to the more structured Victorian English estate gardens that used restrained designs with massed beds of brilliantly colored greenhouse annuals.They are more casual by design,depending on grace and charm rather than grandeur and formal structure.The influential British garden authors and designers,William Robinson at Gravetye Manor in Sussex,and Gertrude Jekyll at Munstead Wood in Surrey,both wrote and gardened in England.Jekyll's series of thematic gardening books emphasized the importance and value of natural plantings were an influence in Europe and the United States.Also influential half a century later was Margery Fish,whose surviving garden at East Lambrook Manor emphasizes,among other things,native plant life and the natural patterns produced by self-spreading and self-seeding.

The earliest cottage gardens were far more practical than modern versions—with an emphasis on vegetables and herbs,along with fruit trees,beehives,and even livestock if land allowed.Flowers were used to fill any spaces in between.Over time,flowers became more dominant.Modern day cottage gardens include countless regional and personal variations of the more traditional English cottage garden.

Kitchen garden or potager

Fig. Formal potager at Villandry,France

The traditional kitchen garden,also known as a potager,is a seasonally used space separate from the rest of the residential garden - the ornamental plants and lawn areas.Most vegetable gardens are still miniature versions of old family farm plots with square or rectangular beds,but the kitchen garden is different not only in its history,but also its design.

The kitchen garden may be a landscape feature that can be the central feature of an ornamental,all-season landscape,but can be little more than a humble vegetable plot.It is a source of herbs,vegetables,fruits,and flowers,but it is also a structured garden space,a design based on repetitive geometric patterns. The kitchen garden has year-round visual appeal and can incorporate permanent perennials or woody plantings around (or among) theannual plants.

Shakespeare garden

A Shakespeare garden is a themed garden that cultivates plants mentioned in the works of William Shakespeare.In English-speaking countries,particularly the United States,these are often public gardens associated with parks,universities,and Shakespeare festivals.Shakespeare gardens are sites of cultural,educational,and romantic interest and can be locations for outdoor weddings. Signs near the plants usually provide relevant quotations.A Shakespeare garden usually includes several dozen species,either in herbaceous profusion or in a geometric layout with boxwood dividers.Typical amenities are walkways and benches and a weather-resistant bust of Shakespeare. Shakespeare gardens may accompany reproductions of Elizabethan architecture. Some Shakespeare gardens also grow species typical of the Elizabethan period but not mentioned in Shakespeare's plays or poetry.

Rock garden

Fig.A naturalistic rockery in England

A rock garden,also known as a rockery or an alpine garden,is a type garden that features extensive use of rocks or stones,along with plants native to rocky or alpine environments.

Rock garden plants tend to be small,both because many of the species are naturally small,and so as not to cover up the rocks.They may be grown in troughs (containers),or in the ground.The plants will usually be types that prefer well-drained soil and less water.

Fig. Rock garden in Chandigarh,India

The usual form of a rock garden is a pile of rocks,large and small,aesthetically arranged,and with small gaps between,where the plants will be rooted.Some rock gardens incorporate bonsai.

Some rock gardens are designed and built to look like natural outcrops of bedrock.Stones are aligned to suggest a bedding plane and plants are often used to conceal the joints between the stones.This type of rock garden was popular in Victorian times,often designed and built by professional landscape architects.The same approach is sometimes used in modern campus or commercial landscaping,but can also be applied in smaller private gardens.

The Japanese rock garden,in the west often referred to as *Zen garden*,is a special kind of rock garden which contains few plants.Rock gardens have become increasingly popular as landscape features in tropical countries such as Thailand.The combination of wet weather and heavy shade trees,along with the use of heavy plastic liners to stop unwanted plant growth,has made this type of arrangement ideal for both residential and commercial gardens due to its easier maintenance and drainage.

East Asian gardens

Japanese and Korean gardens,originally influenced by Chinese gardens,can be found at Buddhist temples and historic sites,private homes,in neighborhood or city parks,and at historical landmarks such as Buddhist temples.Some of the Japanese gardens most famous in the Western world and Japan are gardens in the *karesansui* (rock garden)tradition.The Ryôan-ji temple garden is a well-known example.There are Japanese gardens of various styles,with plantings and often evoking *wabi sabi* simplicity.In Japanese culture,garden-making is a high art,intimately linked to the arts of calligraphy and ink painting.

Contemporary garden

Fig. Contemporary garden

Fig. Contemporary water feature

The contemporary style garden has gained popularity in the UK in the last 10 years.This is partly due to the increase of modern housing with small gardens as well as the cultural shift towards contemporary design.This style of garden can be defined by the use 'clean' design lines,with focus on hard landscaping materials like stone,hardwood,rendered walls.Planting style is bold but simple with the use of drifts of one or two plants that repeat throughout the design.Grasses are a very popular choice for this style of design.Lighting effects also play an integral role in the modern garden.Subtle lighting effects can be achieved with the use of carefully placed low voltage LED lights incorporated into paving and walls.

Residential gardens

A residential or private domestic garden,is the most common form of garden and is in proximity to a residence,such as the 'front garden' or 'back garden'.The front garden may be a formal and semi-public space and so subject to the constraints of convention and local laws.While typically found in the yard of the residence,a garden may also be established on a roof,in an atrium or courtyard,on a balcony,in windowboxes,or on a patio.Residential gardens are typically designed at human scale,as they are most often intended for private use.However,the garden of a great house or a large estate may be larger than a public park. Residential gardens may feature specialized gardens,such as those for exhibiting one particular type of plant,or special features,such as rockery or water features.They are also used for growing herbs and vegetables and are thus an important element of sustainability.

DESIGN PRINCIPLES IN GARDEN-MAKING

Since the space that the garden occupies is three-dimensional,the starting point of design is to get inside that space and create it from within.Imagine being inside a piece of sculpture.The developing space needs to evolve to accommodate the use,comfort and pleasure of its creator.The design elements are then employed to determine the way space will be perceived.All artists-photographers,painters,weavers,sculptors...and yes,garden-makers...use these same principles to create something magical.

A GARDEN ARBOR HELPS DEFINE OUTDOOR SPACE IN BOTH WINTER AND SUMMER

Line is impressed upon all of us from earliest childhood —remember defining objects with connect-the-dots drawings,or the burden of having to carefully color inside the lines? Later one had to learn to write letters on a straight line as well as discovering just what the horizon line meant.In garden design,the form of a line creates a sense of direction as well as a sense of movement.The eye automatically follows a garden line,whether it be the edge of a walkway,the curve of a flower bed,or the outline of plant materials.The character of a line yields specific responses.Gentle,slow curves and horizontal lines tend to be experienced as restful while jagged diagonals or vertical lines create more excitement and tension.

Form,the shape defined by line,is probably the most enduring element in garden design.It is what is seen when first looking at a garden from a distance.Every plant has a distinct growth-habit,a unique mass and volume which develops and changes as the plant matures.

These shapes,whether pyramidal,weeping,columnar,spreading,or round,divide and define the spaces in the garden.Some forms are more dramatic than others and so attract attention.The siting of a specific plant may block a view,or open a sight-line,or alter the view depending on the maturity and growth-habit of the plant—open or compact,herbaceous,evergreen or deciduous.These plant qualities often change with the seasons and restructure the lines of the garden.The form of the plants selected and their placement are critical to creating comfortable,dynamic spaces and pleasing silhouettes.

Texture in the garden creates sensual and visual excitement.It is generally read as the mass and void of foliage,bark,or flowers and changes with the light during the day and with the seasons.Up close,the size and shape of the leaves and twigs become the predominant textural elements of a plant.From a distance,it is the quality of light and shadow on the entire form,the patterns of light and dark,that translates as texture.Rough,coarse textures tend to create an informal mood and are visually dominant,while fine,smooth textures are associated with formal,elegant,subdued moods and are visually more passive.Fine-textured plants are visually translated as being farther away,so fine textures can be a tool for providing a sense of perspective in a small garden and making the space appear larger.On the other hand,the predominance of coarse-textured plants make a garden space appear smaller.Strong textural contrasts add drama and interest to a garden.Bark and foliage are two ways of adding textural interest to any space.Foliage and Spring flowers,with both textural and color interest,are shown in a May garden.........

Scent in a garden is often neglected.Introducing a variety of fragrances will bring an extra dimension to the garden by expanding sensory awareness.If the garden is exposed fragrant plants may need to be located in a sheltered location.The scent of delicately fragrant plants is also more appreciated if they are located near a path or at the edge of a patio or entry area.Specific fragrances,like colors,evoke emotional responses and can help create a certain mood or sense of time in the garden.

Color is often a confusing and puzzling design element for many gardeners.On the other hand it seems to be the one and only element some gardeners consider when planning a garden.Although color is a key element in the design of a garden,many give it too great an importance and fret continually about the often complex rules which some designers have propounded.One of the following three widely employed formulas for planning color in the garden are best used:

1. Design in a green monotone with only an occasional splash of another color,as exemplified in traditional Japanese gardens.
2. Translate from nature,using harmonies of colors,or kaleidoscopic patterns as might be found in a wildflower meadow.

Use the artist's color wheel and paint pictures in the palette-gardening approach made famous by Gertrude Jekyll.

The gardener's final choice of a formula is dictated by location,the size of the garden,and the kind of garden wanted.Living in the countryside just outside of Portland,Oregon,I prefer to translate from nature so that my garden blends in with the natural beauty of the area.But since I also have acres of ground at my disposal I have room for many flower and mixed borders.In designing them I develop specific color schemes using the palette approach,creating for example,a white garden or a blue border.Generally,the more area to be dealt with the more complex the color scheme can be.A garden created in limited space will be more dramatic if the color scheme is kept as simple as possible.

Research has identified the emotional responses specific colors typically generate.The bright reds,yellows,and oranges tend to excite.The softer blues,pinks,greens,and violets produce a calming,tranquil effect.This is one reason why the monochromatic green gardens of the Japanese are so revered.The 'music of the color green' is a phrase often heard in reference to the basic garden color green broken into numerous tones ranging from blue-green to yellow-green.White tends to be the great unifier,providing a neutral,yet somewhat uplifting spirit.Gardeners need to employ an awareness of color responses when planning the functional needs of garden spaces.In general,warm colors -red,yellow,orange- attract the eye,standing out or advancing.Cool colors -blues,most violets and some greens- recede into the landscape.Color therefore contributes to a sense of depth by defining spatial relationships.Remember too that colors in the landscape are not static,changing with the time of day,cloud cover,and season.Color intensity directly relates to the amount of reflected light.Flower color is transient,while foliage,bark,seed pods,and berries provide color highlights and interest at other times of the year,so must loom large in design considerations.

To create a garden space satisfying to the senses and imparting a feeling of unity with the environment,gardeners must also consider six basic principles of design: repetition,variety,balance,emphasis,sequence,and scale.

Repetition is the continuing thread in a garden and is generally defined as duplication.When any design element is repeated the mind is better able to understand the composition as a whole and so a sense of order is introduced..It is the qualities or character of an object—line,form,texture,scent or color- that are usually repeated.Repeating finely textured plants in a garden helps to unify the design and impart a powerful sense of simplicity.Repetition is simply a matter of holding one design quality constant while varying the others.A word of caution: If repetition is carried to extremes the garden will become either monotonous or so subtle that the viewer only sees disorder.

Variety is the life of the garden.The design qualities of line, form, texture, scent, and color are changed and contrasted to provide diversity and avoid uniformity.Diversity develops a tension which helps to hold the observer's attention while creating excitement and enjoyment.Variety is the opposite of repetition.But when it is overdone by adding too many elements,chaos results,so a very fine balance between repetition and variety is needed to achieve unity in a landscape.

Balance refers to the stability or repose of the garden,and is realized by creating an equilibrium between the parts that make up the whole.Line,form,texture,scent,and color all attract our attention so these sensual energies must be gauged and then balanced out.One form of balance relates to layout along a preconceived central axis.That axis can either be informal or formal in its arrangement.Formal or symmetrical arrangements are exactly the same on either side of the axis,while informal or asymmetrical arrangements are unlike on either side of the axis.Another way of conceiving an axis is in the vertical dimension.Natural,informal landscapes which are increasingly popular,depend upon balancing vertical and horizontal elements or small,dense masses balancing large,diffuse groupings.In all cases the elements being balanced must both hold the same importance in the eye.

Emphasis refers to those garden elements which initially seize attention and to which the eye continually returns.It is the creation of the more important and the less important elements in the garden.The parts of any composition should not be equal in their visual interest.Certain parts should be different,perhaps larger,of a contrasting color,form,fragrance,or texture than the rest,depending on the function of the design.Again,if too many elements are introduced the effect is lost.Emphasis can be achieved only by limiting the number of dominant design elements. Sequence is the movement of the garden.It is the rhythms that develop when line,form,texture,and color are changed in a consistent way to lead in a particular direction or to a point of focus.Sequence helps to connect the various design elements.It can be achieved through repetition,being careful to avoid a monotonous repeat; or by progression,such as using textures in graded steps from fine to coarse; or by alternation,a repetition of two or more contrasting features.

Scale within the garden,as distinct from the overall scale of the garden,refers to the harmony of the garden.That is,all the elements of a garden should agree in the sense they convey of the size of the whole.The actual size of an object is different from its relative scale or proportion in relation to other neighboring objects.So scale is concerned with the relationship between the size of an object to the size of the other objects within the same composition.Thus,a tiny alpine plant is out of scale among tall trees,just as it would be planted next to a large building. With these general principles in mind,applied in connection with the elements of line,form,texture,scent,and color,a simple garden space can become a work of art.

Before and After: A Pergola helps define space in a narrow urban backyard

SUSTAINABLE GARDEN DESIGN

PRINCIPLES

1. To design a landscape that minimises the requirement for energy inputs.These inputs may take the form of petrol to run mowers,leaf

blowers and line cutters; chemicals to treat pests; and fertilisers to promote growth,H2O,cleaning agents,stains and finishes to keep hard surfaces clean and well-maintained.Informed plant selection that reduces the need for maintenance inputs – e.g.gardens/landscapes that feature a high proportion of amenity lawn require much higher energy inputs than a mixed herbaceous/shrub planting.On-site treatment of green waste also reduces the need for energy input.

2. To design a landscape that minimizes the requirements for high water inputs,above that which naturally occurs in the particular region.This may be achieved via plant species choices,microclimate design (hydrozoning),mulches,water recycling etc.
3. To design a landscape that maximizes opportunities for biodiversity at all levels.This includes attracting wildlife,maintaining complex ecosystems,companion planting,considering the health of soil biota,recognizing the links between the elements of the garden and the organisms that inhabit it.

4. To design a landscape that maximizes vegetative biomass.This aids in carbon stabilization.For example a landscape that features a high

proportion of paved or hard surfaces and/or high proportion of amenity lawn stores much less carbon than a landscape which features higher proportions of vegetative biomass.And we mean permanent vegetation,not material that must be constantly pruned or mown heavily,or seasonally replanted.

5. To design a landscape that maximizes the opportunity for the growth of produce (vegetables,fruits,bush tucker/edible weeds),and other useful materials including composting,on-site green waste recycling,space for chooks and other sustainable elements.It encourages you to supplement your diet with freshly grown produce,encourages you to consider more than the ornamental value of gardens,and makes you aware of the environmental impacts of broad acre farming and all that this entails,eg.fertiliser/chemical applications,soil structure,etc
6. To design a landscape that minimizes the risk of weed-escapees moving into native habitats.Consider the reproductive biology of the plants selected for your garden,or the ways in which particular species can be maintained to lessen the risk of their unwanted spread (for example,deadheading or removing the flowers off Agapanthus as soon as the flowers die).
7. To design a landscape that minimizes or eliminates the use of materials that disrupt,destroy,pollute or damage natural systems/ communities where they are sourced.For example,retain top-soil so far as possible in present condition,chose mulches sourced from timber industry by-product or local source by-product,avoid sleepers taken from native forest,ornamental river pebbles harvested from active waterways,and avoid plants harvested from the bush or logging coups.Choose locally sourced (eg bulk) materials to reduce product miles,where possible.
8. To design a landscape that minimizes the risk of disruption,pollution or interference to other systems.For example,the effect on non-target areas from highly toxic,mobile or residual chemicals can be catastrophic.Runoff from poorly designed landscapes can affect local systems via erosion or movement of damaging products (chemicals,soil movement,weed seed).

GETTING STARTED

Measuring Up

Any existing house plans or the Land Title will be of assistance,as measurements will be shown on these and can reduce your work.Crudely draw the shape/boundaries of the garden on a sheet of paper.Measure the dimensions of the site and write all the measurements on the paper along the appropriate boundary lines etc.,as shown in the illustration here.Measure the position of existing trees and other plants that will not be removed and note these on the paper.(It's also a good idea to note plants that you are removing,as it can help with planning)

When plotting the position of elements in the garden,utilise existing permanent structures such as fences and buildings,so that measurements can be taken at right angles from them.This will ensure that the position on your plan is accurate.For example,to pinpoint the exact position of the lemon tree on the top right of the illustration here,the tape measure was laid out at right angles from the top fence (the tree is three metres in),and then the tape measure was laid out at right angles from the side fence and that distance is four metres.

Site Analysis

A careful site analysis makes it much easier to design your new garden because you will have noted anything that could have an impact on the garden's success.

a. Where's north? If in doubt,the street directory or the GPS can help and so can a little compass! When designing a garden,the movement of the sun can be very crucial.
b. Take note of sunny and shaded areas.(Remember that the sun is much higher during the summer months.) Local shade will vary depending on surrounding structures eg fences,trees,neighbouring houses,as well as the seasons.

c. If you know your garden well you might even be able to note the areas that are boggy in winter or particularly dry in summer.

d. Slopes.Depending on what work you intend doing or having done in your garden you may need to consider having accurate levels taken,especially for any construction works.
e. Soil type (and condition).
f. Size and position of trees and other plants that are staying.
g. Existing paths and other features that are staying.(These will also need to be measured and accurately noted on your plan).
h. Overlooking to or by neighbours.
i. Views (from windows into the garden as well as views from and within the garden).
j. Neighbours' trees etc.

Take a note of anything that may help you with your garden plan and write it on the paper with the measurements.Remember that even minor ground depressions can be utilised.

Your Needs,Wants and Budget

If you were a professional designer,this stage would be referred to as the Client Brief.In this case you are the client and the designer!

You need to consider what you want out of this garden but keep your budget in mind.If you don't have a lot of money to spend you might want to consider how you can implement works over time.

Needs

Write a list of everything you need in your garden.For example,a garden shed,and the size you are thinking of.It's a good idea to have catalogue information for these sorts of items,so that you have accurate dimensions.

You might need an area for the kid's trampoline and swing set.So you need to work out how much area to set aside – remember to include plenty of surrounding area for sliding off slides,jumping off swings and trampolines etc.

If you are including a vegetable garden,you need to think about the size and whether it will be in ground or above ground (garden beds,pots,or upright planter beds).If you are new to vegetable growing,start small but perhaps have a plan to increase its size should you find you can cope with growing more.Vegetable gardens can be quite time consuming. And because vegetable gardens will require regular water,remember that any other high water use plants you might want to include,such as fruit trees should be planted in the same area.But be mindful of future shading of the vegetable garden.Think about the size of the tree when it has grown,as its canopy might be several metres in diameter. There are smaller fruit tree varieties available now that are small in stature but produce quite a lot of fruit. Then there are the other utility items such as clothes lines and storage areas. Do you need a cricket pitch? And if you have a courtyard sized garden,you better prioritise your needs.We'll get onto some design solutions for smaller spaces.

Wants

This could include the style of the garden.Garden design books and magazines will help you narrow in on the look that appeals to you.

From a design point of view,don't forget to incorporate standard design elements,including repetition,contrast (colour,foliage),asymmetry and surprises in to your overall 'look'. Keep in mind when thinking about what style of garden appeals to you; you can achieve it using drought tolerant,native or indigenous plants.SGA has prepared information showing just how this is possible,and with examples of plants from the local Melbourne area that would be appropriate.

Download a pdf of our Garden Plan examples for some more ideas.

Make a list of your favourite plants,textures,foliage colours and flowers.

Sculpture and artworks might be considered.Even a very small budget can cope with a bowl of water on a rock and this can be as effective in the right surrounds as any expensive sculpture.

Take Your Time

Take your time to think about what you really need and want in your garden.

Some more questions to get you thinking...

Do you like winding paths and surprises around corners,or do you prefer a vista of a garden? Do you entertain outdoors a lot? If so,you may need to consider an entertaining space as a need.

Do you have pets and how might you cater for their needs,especially rambunctious dogs and young plants? Are there any other considerations relating to the wider landscape that you might need to consider,such as bushfire,heritage and tree overlays? If you are considering built structures,contact your local council to find out their requirements regarding permits etc.And while you're at it,ask for their list of garden weed plants in your area. Movement throughout the space is a key aspect.Movement is not only important for navigation around the space,but should be factored in (eg the placement of paving stones) so that compaction of soil can be avoided,which affects soil drainage and nutrient uptake by plants.

THE DESIGN

We will now adjourn to the drawing table (a la kitchen table),with crude hand drawn plan with site measurements written on it,a site analysis,a firm idea of wants and needs,and a firm idea of limitations budget-wise.

Scale

All drawing from now on should be to scale.This means that the measurements on the paper must reflect actual membership but smaller scale.In landscape drawing we often work in the 1:100 (which means that 1 cm on the paper is equal to 100 centimetre on the ground.This is the same as saying the 1 cm on the paper is equal to 1 metre on the ground).1:50 is also a common scale.This means that 1 cm on the paper is equal to 50 cm (or half a metre) on the ground.

A 1:50 scale is most often the best,as the plan can provide enough detail at this scale.For larger properties,many different drawings may need to be produced,including Shadow Diagrams and Section Elevations. Scale rulers are available at office suppliers.

Paper

Tracing paper is good to work on,as you can put another sheet on top and quickly rework if necessary.Often it is available by the sheet in larger sizes too.The size of your garden and the scale you want to show will dictate the size of the paper!

What is the longest measurement on your rough drawn plan – maybe it's the measurement from one end of the garden to the other? If it's 15metres that will be 30 cm on your paper if it's at the 1:50 scale.

Other tools

There is a multitude of professional tools available but for the one-off (or maybe a couple-off) home design plans most of them can be avoided.A scale ruler can be handy but is not a necessity,cheaper drawing pens are quite suitable,and a template of scaled circles (as shown here) is mighty useful for drawing plants to the right diameter.

An H pencil,which is quite fine,is good for drawing on the tracing paper,whereas an HB or heavier leads 2B and 4B are good for sketching (this depends on personal preference too).

A bendable curve (also shown here) is useful for shaping curves. And a triangle will help with right angles. A drawing table like this portable one costs about $180,but even a smooth,clean and dry kitchen table will do the job.

Drawing

All those rough drawn lines with the measurements on them,must now be transferred accurately to your paper. This can be time consuming,depending on the level of detail and size of your garden.Draw up all the basic building and boundary measurements first. Next,incorporate the elements you are keeping from your existing garden onto your new plan. Trees and other elements that are staying must be put in the right spot and shown at the right size and dimensions.For existing trees,this means that your circle template reflects the actual spread of the canopy – or rather,the diameter of the canopy.This is important to help you gauge shading effects when placing your new plants and elements. Keep your scale in mind when drawing.For example,when you are drawing your new deck,go outside and measure your new dimensions on the ground too and actually see the size of it.

Paths

Paths need not be all the one size,but consider how they are to be used.For example,you don't want to make the access path to the shed very narrow as you need to be able to negotiate it with a wheelbarrow and tools spilling out.

Placing stuff

Consider the view out of your windows and what you will be looking at.You don't want to spoil a potentially appealing view by sticking the new garden shed right in the way.And the same goes for artworks or focal points too.Think about where they will be viewed from – out of windows and when walking around the garden.

Consider utility

Consider how quickly and easily you need to get from one point to another.Sometimes we don't just want to meander around,sometimes we are out there working,and easy access and speed can be a necessity.

Selecting Plants

At this stage think about the size of the plant and the look you want for a particular location,rather than the specific plant.For example,you have decided that along a path's edge a groundcover that has yellow flowers most of the year would be ideal.In another area you are thinking that a shrub with grey foliage that will grow to a maximum of 2 metres will look great next to the existing native hibiscus (*Alyogyne huegelii*) and its bright purple-blue flowers.You can write this briefly on your plan in light pencil to help remember what you have in mind.

When drawing in plan view,all you can show is the diameter of the plant,not its height.Using a thicker pen on the final drawing gives the illusion of a more dominant (bigger) plant.The diameter that's shown on the plan is the plant's diameter at maturity,not when it was first bought in the pot!

Special Elements

Adding a Rain Garden or designing a pond is one thing,but building them is quite another.Keep in mind your budget and capabilities when designing these sort of elements and structures.You may need to consider expert help,or even a builder for some of the built structures.Local councils have lots of information of this kind.

Plant Selection

A plant's suitability for the job,the site and the position,is the rule for all plant selection in sustainable design.By this we mean that the plant will do what you want it to do (for example,provide shade or screening),its suitability for the climate and microclimate (for example,roses might be very suitable for your climate,but not if they will be sitting under full

shade all summer long),and,most importantly,the soil.

Not only does wise plant selection make life easier from a maintenance point of view but it is a key to true sustainability.After all,many plants die simply because they are placed in the wrong spot or are not suited to the local soil characteristics (sandy,loamy,clay or combination soil).

In recent times,drought tolerance has been a key criteria for plant selection,and there's no signs that this will change soon in (change to) various regions. Weather patterns are a lot less predictable,nowadays.It's becoming equally important to select plants that tolerate extreme conditions,ie.heavy rain,high winds,and severe heat,as well as drought. This need not impact on the style of garden,as plants tolerating extreme conditions can be found to suit all garden styles.

Weeds

An important point to remember is to avoid environmental weeds.Your local council should be able to provide you with a list of weeds that are a problem in your region.Otherwise,Google is a great resource! If you have any doubt,simply type the botanic name of the plant and check out Australian websites.If it's a weed,it will come up on sites such as Weeds CRC,council sites and many others. Another important issue is the state of your soil,especially if you are going to be creating your garden on a recent building site.If the topsoil has been lost and you are left with subsurface clay and the like,you will need to improve your soil well before planting.Recycled green waste compost and soil improvers are available in bulk and should be added to your soil months ahead of planting.This will allow time for worms and soil microflora to recolonise and create a more suitable planting environment.

Avoid importing topsoil,as this is usually from an unsustainable source.Try to work with the top-soil on site,as this can always be ameliorated and is teeming with important organisms (biota).A useful book on the subject of soils is Gardening Down-Under by Kevin Handreck (published by LandLinks).

Local Plants

Indigenous plants are plants that grow naturally in your area.They have evolved with the local climate and the soil,so are usually perfectly suited to your environment.However,if you have a recent building site and are looking at subsurface clay etc,as mentioned above,you will still need to improve your soil prior to planting.

There are many advantages to using local plants besides their ability to thrive in your garden.They are often low maintenance and thrive without the addition of fertilisers or pesticides and they provide food and shelter for native wildlife.There are indigenous plants to suit any style of garden,so you don't have to have a bush garden if that doesn't appeal to you.Do not remove indigenous plants from parks and bushland.Your local council should be able to provide you with information on nurseries that grow indigenous plants.

Native Plants

Native plants are Australian plants that aren't necessarily from your area.Native plants can offer similar benefits to your garden as indigenous plants and once again,don't necessarily dictate a bushland garden style.The book,The New Native Garden – Designing with Australian Plants by Paul Urquhart (published by New Holland Publishers) offers great suggestions on using native plants in any style of garden.

Do be careful,though,some native plants can be environmental weeds in your area.For example,Sydney Cootamundra Wattle (*Acacia baileyana*) is a weed in Victoria,and the beautiful Western Australian Blue Bell Creeper *Sollya heterophylla* is a weed in the eastern states (although Austraflora has developed a sterile variety that is not a weed – *Sollya heterophylla x parviflora* 'Edna Walling Blue Bells',shown here).

There is also information on the website of the Australian Native Plants Society. Don't assume,though,that because a plant is indigenous or native,it will be drought tolerant.Some are from riparian (waterway) environments,so are used to having ample available water.*Indigofera australis* (shown here),is an example.It is an elegant small shrub that is indigenous to all States of Australia.It is found naturally mainly in riparian environments and damp forests.

Exotic Plants

Exotic plants are plants from other countries.There are many suitable exotic plants that are drought tolerant species or are suitable for your site,but do check that the plant selected is not a weed in your area.

Mix and Match

There is no reason why you can't incorporate a mixture of indigenous,native and exotic plants in your garden.Obviously if you are including a vegetable garden and fruit trees,then exotic plants will be necessary (although there is also the possibility of exploring bush tucker!).

PLANT PLACEMENT – PLANT STACKING AND HYDRO-ZONING

Be careful when placing plants that,over time,you won't be shading areas that you don't want shaded.This is especially important with regard to the northern aspect,as the sun is lower in the sky in winter and allowing winter sunshine into the house or corners of the garden is very desirable.

The other side of that is creating shade where it is required.Providing shade to western exposures is a good idea,so think about taller trees and shrubs here (where possible).Deciduous plants can be planted closer to the house,whereas evergreen trees need to be placed well away,to avoid casting shade in winter.

Consider competition between plants over time too (for sun,water,ventilation).You need to allow plenty of room between large trees and shrubs.Consider the size of the plant at maturity when drawing them on your plan. Another factor is to incorporate plant-stacking where larger trees or plants can create shade and wind-protection for shade-loving species.Similarly,plants can be zoned according to their water needs,so water isn't wasted.Plants needing more H2O can be placed 'down-slope'.

Also be mindful that you won't be creating a future problem because you are placing a tree too close to a building or to drains.There are web-sites that will list trees with invasive roots,in your local area.And consider your neighbours too.Don't plant large trees close to boundaries,without your neighbour's consent,for example.

SITE DETAILS

Not only can the microclimate vary from one corner of the garden to another,but slope and proximity to buildings can also have a significant impact on plant success.For example,the south side of buildings is shaded for most of the year,and the western side can be shaded for most of the day and then suddenly be blasted by hot afternoon sun during summer.

Soil conditions under existing trees and under house eaves can be quite dry.Slope and depressions can also have an impact.Depressions will often be wetter most of the year,and the top of a slope is generally drier than the bottom.

Grow What Where (which also comes with a CD),by Natalie Peate,Gwenda Macdonald & Alice Talbot (published by Bloomings Books),is an excellent resource for selecting native plants for specific environments.

THE SOFT LANDSCAPE

The soft landscape includes plants,mulches and composts etc.Plant selection guides what soil preparation is needed and a lot of what else is required too.Many indigenous plants require little ground preparation,yet a vegetable garden requires considerable preparation,especially in poor soils.Vegetable gardens are often best built above the soil,using the No Dig Method of establishment and growing.

Indigenous and native plants benefit from a coarse mulch.Some prefer gravel mulches,or light loose leaf litter,so ensure you know the requirements of the plants you are putting in.

Most soils will benefit from the addition of compost and for plants where soil improvement is necessary,it would be wise to consider adding the compost weeks or even months ahead of planting.

There's more information on soil preparation in SGA's Information Pages,under Soil Health.

What you now need to work out is just how much of everything you need.For those of you who have forgotten your school maths,the following formulas are the ones of most importance! Working out volumes of mulch and the like becomes very easy (especially with a calculator!).

Area of a square (and rectangle): l x w (l = length,w = width)

Area of a circle: π x r2 (π = pi,which = 3.141-if you have an electronic calculator, π will be on it as the symbol,so you don't need to type in the actual numbers, r = radius)

Area of a triangle: (b x h)/2 (b = base,h = height)

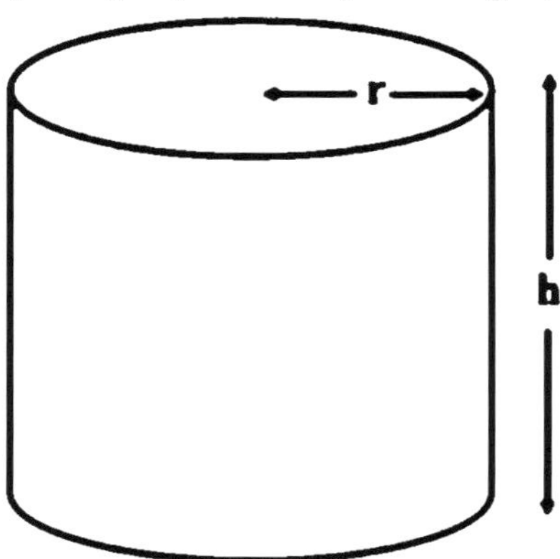

If you need more of an explanation of these formulas,a good webpage is: http://math.about.com/

You can use a combination of mostly squares,rectangles and circles to work out the area of garden beds on your design.

These formulas only give you the area,the other variable is depth.And depth depends on what you need.For example,to work out how much mulch to order for garden beds,work out the area of all the beds.Now,we usually like mulch to be up to 70mm in depth,so we multiply the area by the depth of 70mm.Be careful with units and decimal points.

For example,if all the areas of garden bed give a total of about 180m2,then the volume of mulch required will be 180 x

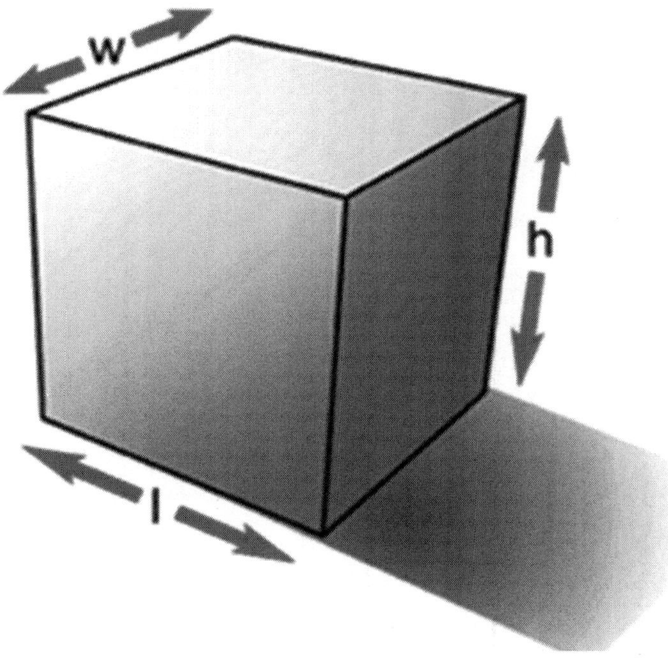

.07 (notice the decimal point has moved because we are working in units of metres).The total is 12.6 cubic metres (m3),but when ordering you would probably go up to 13 or maybe even 15 – to round it off (many of us often add another cubic metre or so for good measure!).

You will need to go through the same procedure for all materials.

And another couple of formulas that are very useful are the volume of a cylinder and the volume of a cube.Both or either will be useful if you have included a lot of large pots in the design,as you will be able to work out how much potting media to purchase.

Volume of a cylinder: ð x r2 x h

Volume of a cube: l x w x h

The Hard Landscape

The hard landscape is essentially anything that's not the soft landscape! The soft landscape includes the plants,mulch

and composts etc.

We're left with decking,paving,walls,and essentially the built components of a landscape.This can be a tricky area for the inexperienced or untrained,and there are some very important issues to consider before you tackle building structures or hard landscaping.

Material Selection

Selecting the materials you want can be time consuming alone,with the amount of product available out there.But to try to make an informed decision based on environmental criteria,adds a whole other dimension!

Here's a guide:

- Where does the material come from? Local products require less transport,chain of custody/forest stewardship is easier to establish,and local products have less impact on greenhouse gas emissions and other negative environmental impacts.
- Avoid rainforest timber.(There is always a sustainable replacement.)
- Use plantation grown timber or bamboo from a sustainable source.
- Look for radially sawn timber,as this technique produces more timber per log.
- Consider recycled products where possible (eg reclaimed bricks.)
- Consider recycling existing material.(For example,concrete can be sent to a recycler and turned into crushed rock which can be reused on site.Crushed bricks can be used as back-fill for walls.)

- Where do those pebbles come from? Are they mined without control in Asia? Or are they at least mined from a controlled source locally.
- Consider the urban heat island effect and wind-turbulence factor.The greater the proportion of hard surfaces you build,the lower the heat absorption by vegetative bio-mass that can occur and the harsher the wind-factor may be.Also,by reducing area for vegetation,you reduce the potential for ground-water cleaning,stormwater filtration and carbon uptake,as well.
- Factor in the amount of hard surface cleaning and maintenance you wish to do.Outdoor walling and decking can require regular cleaning,sanding and staining.
- Finally,consider allowing for permeable areas such as crazy paving/ cobble-stones/stepping stones with space for vegetation,rather than an en masse area of non-permeable paving.

Know When NOT to DIY

Know your limitations! Simple paving may be well within most people's scope of ability but unless you have training in bricklaying,paving,etc,think seriously about the outcome.

The hard landscape is the most expensive element of a landscape.It's also the most obvious,so a shabby job will permanently detract from the overall look.

There are safety issues inherent in built elements too.We've all heard horror stories of DIY walls falling over.

Training is obviously an option.There are plenty of short courses available through TAFE and the like,which will ensure you develop enough technique to create the more modest aspects of hard landscapes.

There are times,though,when it is compulsory to bring in professionals,so be mindful of building regulations.Contact your local Council for information on their requirements,rules and regulations.

Be mindful of easements on your property,as it is illegal to build over easements.

Dial Before You Dig (1100) is an essential service to contact prior to any major excavations.Dial Before You Dig advise as to where underground services are located on your property.

Costs

Even if you don't feel confident about building the hard landscape,there is research you can undertake that will help you keep an eye on costs.

Ascertaining the amount of paving and decking required uses the same formulae as used to estimate soft landscape materials,except that you don't need to work out depth! Suppliers can often help you work out the costs of decking and paving based simply on the area.Get several quotes for materials,as this will help to ensure you are getting the best price.

Water Features

Water features in a sustainable garden are ideally designed and constructed to increase biodiversity.Attracting frogs to

your garden is an especially satisfying reward for creating suitable habitat.

Frogs are not only interesting critters to have around but they also devour huge quantities of mosquitoes,flies and other insects.

There are a number of pre-made fibreglass ponds that can be simply dug into place.Edges and base can be hidden and the pond softened with the inclusion of rocks and pebbles.Another simple method of construction is to excavate a suitable sized hole and lay pond liner in it.The edges and base are also hidden with rocks and pebbles.

When designing for a pond,it's a good idea to consider utilising the natural slope of your block,wherever possible.Although you may wish to compromise that if you particularly want your pond in a specific place in your garden.

Placed in lower areas of your garden,the pond may be fed naturally by runoff rainwater,but the other advantage is that the surrounding ground will stay damp.This makes perfect frog habitat because Australian frogs don't live in water all the time.In fact,they are used to their watery habitat drying out during summer and so take to the shelter of lush grasses and plants.

Pond Features

A frog pond can incorporate one or all of the requirements for each part of the frogs' lifecycle.

- Damp bog zone for adult frogs.
- Shallow water zone for laying eggs.
- Deep zone of at least 300mm for tadpoles.

Your frog garden should also have:

- Soft,thick vegetation that droops into the water,for shelter and protection
- Rocks,logs,bark and leaf litter
- Mostly shade

- Sloping sides for frogs to crawl out
- Been made from non-toxic materials (concrete ponds will need to be sealed and plastic ponds be made of food-grade plastic)
- Food plants for tadpoles (and they will eat them,so don't put your prize waterlily in there!)

Frog-Friendly Plants

Frog-friendly plants include the following:

Grasses: Kangaroo Grass (Themeda triandra),Weeping Grass (Microleana stipoides) or Wallaby Grass(Austrodanthonia spp.)

Tufting plants: Kangaroo Paw (*Anigozanthus spp.*) or Black-anther Flax-lily (*Dianella revoluta*)

Bog plants: Soft Water Fern (*Blechnum minus*),Thatch Saw-sedge (*Gahnia radula*),Knobby Club-Rush (*Ficinia nodosa*),Grassy Mat-rush (*Lomandra confertifolia*) and Tassel Cord Rush (*Restio tetraphyllus*) – this is also suitable for planting in water.

Plants for Ponds

There are lots of native water plants that are suitable for garden ponds.These include Villarsia (Marsh Flower) and

Marsilea drummondii (Nardoo).The following books also contain information on Australian water plants:

- Aquatic and Wetland Plants,by Nick Romanowski,1998,published by the University of New South Wales Press (UNSWP).
- Australian Native Plants,by John W.Wrigley and Murray Fagg,1996,published by Reed New Holland.

There are many exotic water plants that are terrible environmental weeds,so be very discerning when you buy,and do a bit of research.

Other commonly used water plants include: Purple Loosestrife (*Lythrum salicaria*),Tassel Sedge (*Carex fascicularis*),Jointed Twig-rush (*Baumea articulata*) and Water Ribbons (*Triglochin procerum*).

Things to Avoid

- Most fish will eat tadpoles.
- Tadpoles and eggs can be killed by fountain pumps.
- Cats and dogs will often hunt and kill frogs.Protect the frog area of

your garden with sharp,spiky plants.

- Pesticides and herbicides.Frogs eat insects,so you don't want to spray them.And frogs are very sensitive to chemicals,so you can't have both.
- Fertiliser runoff.
- Allowing duckweed or *Azolla* to cover the top of the pond as it reduces the oxygen available to tadpoles.
- Cleaning out the pond too often.Tadpoles need some material to be breaking down in the pond water to provide food for them.
- Collecting tadpoles from the wild.It is illegal in most parts of Australia.It's also unnecessary.If you provide the right habitat,they will find you!

TEN ELEMENTS OF NATURAL DESIGN

The elements that make a landscape design "natural" are difficult to define.A landscape with curved bed lines,informal plant arrangements and no pyramidal yews does not always qualify as a natural landscape.And advocates of natural design are not necessarily eager to banish a host of beautiful exotics from the plant palettes of American landscape designers,replacing the plants with a motley crew of straggly natives. The basic concept behind natural design,however,is fairly simple—to incorporate native plant communities into the designed landscape.But their successful incorporation requires a basic understanding of how native plants operate in nature.

Too often,random informality passes for "natural," when in reality nature is highly ordered and anything but random.Understanding this order and using it in our designs is the key to making natural design workable and successful.This does not mean,however,that we must design exclusively with native plants,attempt to copy nature exactly,or exclude the influences of other design styles.The goal is to create a framework for the overall designed landscape that has an aesthetic and ecological relationship to our indigenous landscape through the use of native plants in their natural associations.

Fig. The basic considerations of natural design can be broken down into three categories: aesthetic,managerial and environmental.

The aesthetic aspect of our designs is highly subjective,and individual style varies greatly.Some designers may object to uniformly patterning their work on the native landscape,feeling they are homogenizing their designs or stifling their artistic expression.But,as landscape designers,our medium is the land.

Unlike a painter whose art occupies an isolated canvas,our work visually interacts with the surrounding landscape,both natural and constructed.We therefore have a responsibility to contribute continuity and a sense of place to the larger landscape.To successfully accomplish a marriage of art and nature,we should sometimes put our egos aside and let nature be our guide.

The managerial aspect of natural design is tied to the fact that reducing landscape maintenance is a strong priority for virtually all our clients.Natural design techniques can make a great contribution in this regard.This does not mean that natural landscapes are maintenance-free and can be completely left to natural processes with no human guidance,however.

Fig.: Rob Cardillo

What natural design does mean is that landscapes that incorporate native plants and natural processes will require less time,money and energy for upkeep than designs in which plants are selected and combined for ornamental effect alone.A purely ornamental garden is like a beautiful,sleek automobile with no engine.It may be nice to look at,but the only direction it will go without help is downhill.We will be perpetually required to tow these gardens up the hill with fertilizers,watering hoses and weeding forks.

The environmental considerations of natural design are equally important.Many detrimental landscape practices can be minimized or eliminated.Such landscape practices include the excessive use of pesticides,herbicides,inorganic fertilizers,fossil fuels burned while mowing large areas of turf grass,and exotic species that have aggressively naturalized in the wild.

Natural design aims not only to reduce these negative effects,but to make a positive contribution to the surrounding environment as well.Naturally designed landscapes can also become functioning ecosystems capable of providing food and shelter for animals and insects,while helping to perpetuate many native plants whose habitats are being reduced through development.

Cultivate in your clients an appreciation of the beauty in nature.

Everyone admires the beauty in a majestic mountain range or a towering waterfall,but most of what we can create in our landscapes is more subtle.The contrasting patterns of straight and leaning tree trunks in a woodland grove,a single turk's cap lily (*Lilium superbum*) nodding above a bed of meadow grass,or the layered branches of a pagoda dogwood (*Cornus alternifolia*) in a woodland edge may be an acquired taste.

A native old field in winter is a prime example of how learning to see the landscape anew can open a whole new vista of aesthetic possibilities.The glistening orange of little bluestem (*Schizachyrium scoparium*) in the sun,punctuated with columnar green patches of eastern red cedar (Juniperus virginiana) is a spectacular American scene,and a much more warming sight on a frigid February morning than a curled up 'PJM' rhododendron (Rhododendron 'PJM') in a crispy bed of pachysandra (*Pachysandra terminalis*).Designers who cultivate in their clients an appreciation of the natural world around them will find their work to be more easily accepted.

Minimize disturbance of existing native growth.

Protecting existing native growth,particularly woodlands,is easier and less expensive than trying to restore it after it's destroyed.Even our best restoration efforts may never achieve the beauty and mystery of an undisturbed woodland.Developers,architects and clients need to be aware of the benefits of considering ecological systems before designing the structures for the site.Early decisions relating to the siting of buildings,topographic changes and excavation disturbance can help minimize destruction of natural growth during construction.Unfortunately,landscape designers and architects often are brought in after construction is complete and have no opportunity to influence the treatment of the existing landscape.

Decide how closely your design will emulate the native landscape.

The design will be determined by numerous factors including the character of the surrounding landscape,client dictates,architectural style,site characteristics and the scale of the site.A large site may allow for the design of a functioning ecosystem using strictly native species.A smaller residential site can be designed with a perimeter of site-appropriate natives,becoming more cultivated as the landscape nears the house.Native plant cultivars such as 'Golden fleece' goldenrod (*Solidago sphacelata* 'Golden fleece'),'Purple Dome' aster (*Aster novae-anglia* 'Purple Dome') and native azalea cultivars can be very useful in making a transition from wild areas to more formal ones.

Allocate the location of woodlands,open spaces and transitional areas.

Natural landcape patterns found in many areas throughout the country are formed by the interplay of woodlands,open landscapes and the transitional areas where they meet (edges or ecotones).A graceful and functional mix of these features will define the design before any plants are selected.Even small properties can be approached in this manner,often resulting in the illusion of more space.

Base your design on native plant communities found in similar conditions in the surrounding areas.

Determine which plant communities would have existed on the site had it not been disturbed,and use these as a design model. Determining native plants is easiest on a site that still contains remnants of indigenous growth.If this is not the case,you can obtain information by observing nearby natural areas with similar ecological conditions,analyzing the soil and hydrology of the site,obtaining geological maps and studying the natural history of the area.If the post-disturbance soil and water conditions are no longer capable of supporting these plant communities,consider basing your design on a community with similar conditions.

Use and plan for natural processes of change to modify the landscape.

The indigenous landscape is a constantly changing system composed of plants,animals,insects,microorganisms and soils.Plants are not isolated entities,but participants in a system constantly in flux.Different types of systems change at different rates.The annual meadow immediately resulting from a disturbance may last for only one year,while the perennial meadow may last for 10 before yielding to pioneer forest species.By contrast,an old oak and hickory forest may last for hundreds of years if left undisturbed.

Fig. Rob Cardillo

Once these changing systems are understood,the designer can decide which aspects to encourage,discourage or manipulate to fit the requirements of the client and site.Designed landscapes need not be static photographs frozen in time forever,doing battle with the forces of nature.

Occupy all the spaces.

A basic law of almost any native ecosystem is that if nothing is currently growing in a given space,something soon will.The more available space is filled,the less opportunity there is for a weed to enter.Plants grow against each other,above each other and below each other.Even a 3-foot-tall meadow has a multi-layered structure designed to seal off the area.

This is also evident below ground,where fibrous rooted plants occupy the soil surface and coexist with deep taprooted plants "holding down the fort" down below.There are obvious lessons here for the designer interested in creating landscapes that have the ability to fight off weed invasion without the aid of mulches,fabrics and grub hoes.Mulched beds around isolated groupings of shrubs are an open invitation to neighborhood bullies such as Canadian thistle (*Cirsium arvense*),knotweed (*Polygonum*) and nut grass (*Cyperus esculentus*).

A mixed,densely planted herbaceous ground cover layer,composed of plants with complementary aboveground and belowground growth habits,will be far more successful at inhibiting weed invasion than any mulch.If this ground layer is also designed for succession of bloom and contrasting foliage texture,we can create a reduced-maintenance landscape that suggests the diverse tapestry of our native ground covers while achieving an artistic and colorful composition.

Increase ground water recharge by preserving rainwater on-site.

Current landscape practice often considers surface water as something to be eliminated.Meanwhile,water shortages are a frequent problem in our communities.

Whenever we grade a property to direct surface runoff into the storm water system,we are sending a valuable commodity out to sea.Aquifer recharge,the replenishment of our underground water tables,depends upon the absorption of rainwater into the ground.We can assist this process by using ponds,irrigation catchments,porous paving surfaces and bog gardens.

Low wet areas can be converted into colorful assets by designing them as wet basins containing a range of colorful water tolerant plants like turtlehead (*Chelone lyonii*),Joe-Pye weed (*Eupatorium purpureum*),New England aster (*Aster novaeanglia*) and blue flag iris (*Iris versicolor*).

Employ alternatives to high-maintenance lawns.

The American lawn has become the focus of a great deal of controversy.Great quantities of water,fertilizers and fossil fuels are expended for lawn upkeep and the amount of pollution from herbicides,pesticides and small engine exhaust is well documented.

Although there is nothing inherently evil in a blade of Kentucky blue grass or the person who likes it,replacing substantial portions of mowed lawn with other,more ecologically friendly plantings would have a positive effect on our environment.

A mowed lawn does serve a unique function in that you can walk,lay and play catch on it—activities that are difficult in a tall grass meadow or a cottage garden.It is possible,however,to offer alternatives that are affordable,easily sustainable,ecologically sound and aesthetically pleasing.

The first alternative to lawn is lawn.Not the resource-intensive grass monoculture that we normally plant,but a diverse ground cover of creeping broadleaf plants combined with slow-growing drought and disease-resistant grass cultivars or native grass species.These plants could include buffalo grass (*Buchloe dactyloides*),Pennsylvania sedge (*Carex pensylvanica*),wild strawberry (*Fragaria* spp.) and violets (*Viola* spp.).A lawn of this type would require little or no fertilizer or chemical application,and would need to be mowed less frequently than a traditional lawn.

Wildlfower meadows are currently the most popular lawn alternative as they can provide visually stimulating,low-maintenance landscapes.However,in order for these plantings to succeed in the long run,the majority of wildflower seed producers must completely revamp their mixes.Annuals and short-lived perennials selected for immediate floral effect must give way to long-term native perennials and grasses selected for function and site-adaptability,as well as aesthetics.By patterning these landscapes after our native prairies and grasslands,their exciting potential can be fully realized.

The most neglected lawn alternative is woodland.While open space is highly valued,it can be even more appreciated when contrasted with a shady tree grove.While this type of landscape would obviously take far longer to mature,a transitional period can be filled with a meadow or grassland landscape supplemented with trees.Woodland understory and ground layer plants can be added after a sufficient canopy is developed.

Exclude invasive,exotic plants in the native landscape.

A number of exotic species have naturalized so aggressively into our woods,meadows and wetlands that the natural plant diversity of these areas is destroyed.These include many commonly used ornamental plants such as

Norway maple (*Acer platanoides*),burning bush euonymus (*Euonymus atlatus*),privet (*Ligustrum*)Japanese barberry (*Berberis thunbergii*).Russian olive (*Elaeagnus angustifolia*) and tatarian honeysuckle (*Lonicera tatarica*).Purple loosestrife (*Lythrum salicaria*),a European perennial that has attained enormous popularity,has completely destroyed the biodiversity of thousands of acres of wetlands.(Claims that its cultivars are sterile and therefore harmless have been proved false,as these cultivars eventually hybridize into fertile forms.) We should completely abandon using any plants that have proved to be invasive in the native landscape.

In addition,we should be looking into ways to identify and discontinue using any new plants that show likely potential for invading our natural areas.

Although natural design is not new,current public interest in natural aesthetics,reduced landscape management and environmental issues is making its widespread acceptance a real possibility.In order to capitalize on this opportunity,we need to develop concrete and reliable strategies for the design,implementation and management of these landscapes based on real ecological principles.

Landscape designers and architects influence the treatment of vast areas of land.We have a responsibility to treat the land as more than our personal paint canvas.

The landscape designer should be part artist and part repairman,restoring some of the aesthetic qualities and environmental functions of the native landscape that have been destroyed.By making an effort to truly understand the workings of our indigenous landscape,and combining that understanding with the horticultural and design knowledge long associated with our profession,we can legitimately lay claim to the word "natural" when describing our work.

Woodlands,Meadows and Transition Zones

Natural design can incorporate native woodlands and meadows,as well as transition zones between the two.In a limited space,you can adapt these elements to a smaller scale.Woodlands are the dominant plant community type in many areas throughout the United States.If left undisturbed,these open sites would revert to some type of forest community after passing through various stages of herbaceous and woody shrub composition.Therefore,where woodlands predominate,landscapes patterned after our indigenous forest should be a strong—if not dominant—component of our work.The re-establishment of woodland landscapes on open sites,both large and small,should also be considered as a primary option.

A natural woodland is composed of a multilayered tapestry of canopy and understory trees with a ground layer of shrubs and herbaceous plants.Trees of the same species are found at many different ages and irregular clusters,and do not necessarily have straight trunks and uniform heads.In fact,leaning and jagged trunks can often be the most interesting feature in the landscape.

Unfortunately,there is not enough space to create true self-sustaining woodlands on most residential properties.However,if the planting of site-appropriate native woodland species on the perimeters of suburban properties became commonplace,a series of continuous woodland corridors would be created,connecting existing isolated fragments of native forest badly in need of ecological interaction.The positive ecological impact of this would be quite significant,while the aesthetic advantages of suburban landscape in visual harmony with our native American forest would be easily apparent.Additional privacy would be a bonus.

Native meadows and grasslands are often the product of the disturbance of existing woodlands.This disturbance can result from either people (fire,bulldozers,chain saws) or nature (fire,storm,pathogens).If left alone,a succession of plant communities occupies the site.The process begins with herbaceous annuals and biennials,then leads to herbaceous perennials and grasses.

A mixed old field composed of herbaceous perennials,shrubs and fast growing pioneer trees follows.Finally,the site reverts to woodlands until another disturbance comes along,starting the process all over again.To preserve the open landscape then—be it lawn,meadow or perennial border—we must continuously arrest the successional process by artificially disturbing the landscape in various ways,such as by mowing lawns and weeding perennial gardens.

Although there are many different types of native meadows,they generally have one thing in common.The plant group most vital for their stability are the warm-season grasses such as Little Bluestem (*Schizachyrium scoparium*),Indian grass (*Sorghastrum nutans*) and Panic grass (*Panicum virgatum*).Although wildflowers receive the bulk of attention and are certainly an important aesthetically pleasing portion of the mix,it is the grasses that provide the stability for successful long-term results.Only through a combination of warm-season grasses and tough native perennials selected for site adaptability can we create dynamic and colorful landscapes that can live up to the low-maintenance expectations surrounding the wildflower meadow.

Woodland edges (or ecotones,in the language of the ecologist) are the transitional areas between the woods and the open landscape.They are,by nature,rapidly changing plant communities composed largely of herbaceous perennials,woody shrubs and vines.These communities can include species from both the woodlands and open landscapes that the ecotones separate.Additional species not found in either of the bordering landscapes may also be present.Some of the plants found growing in edges are the same pioneer species found in landscapes in transition from open to woodland communities.

We can design and manage a small woodland as an edge ecosystem,or selectively remove some of the edge species and manage the landscape to contain some of the aesthetic and functional characteristics of an interior forest.Ecotones are dynamic and diverse communities that hold a very prominent visual position in the landscape,and the design opportunities are exciting.Woodland edges also present both challenges and opportunities for the designer.Many of the most pernicious weedy vines such as Oriental bittersweet (*Celastrus orbiculatus*) and Japanese honeysuckle (*Lonicera japonica*) can be found in this type of environment where a combination of trees for support and adequate light for growth are present.A dense planting of desirable edge species,such as Arrowwood (*Viburnum dentatum*),Pagoda dogwood (*Cornus alternifolioa*) and Sweet fern (*Comptonia peregrina*) can help to eliminate these vines,but selective removal may be needed as a supplemental management tool in many cases.

Natural recruitment of *Maiainthemum canadense*,*Dennstaedtia punctilobula*,and *Carex pensylvanica* in a woodland setting.Invasive species have been selectively controlled (through cutting or herbicide treatment) to give the desired species a competitive edge and eventual domination of available resources.The shrub and vine layers can be added with protection from deer browse as necessary.

COTTAGE GARDEN

Fig. Roses,clematis,a thatched roof: a cottage garden in Brittany.

The cottage garden is a distinct style of garden that uses an informal design,traditional materials,dense plantings,and a mixture of ornamental and edible plants.English in origin,the cottage garden depends on grace and charm rather than grandeur and formal structure.Homely and functional gardens connected to working-class cottages go back several centuries,but their reinvention in stylised versions grew in 1870s England,in reaction to the more structured and rigorously maintained English estate gardens that used formal designs and mass plantings of brilliant greenhouse annuals.

The earliest cottage gardens were more practical than their modern descendants — with an emphasis on vegetables and herbs,along with some fruit trees,perhaps a beehive,and even livestock.Flowers were used to fill any spaces in between.Over time,flowers became more dominant.The traditional cottage garden was usually enclosed,perhaps with a rose-bowered gateway.Flowers common to early cottage gardens included traditional florist's flowers,such as primroses and violets,along with flowers chosen for household use,such as calendula and various herbs.Others were the old-fashioned roses that bloomed once a year with rich scents,simple flowers like daisies,and flowering herbs.Over time,even large estate gardens had sections they called "cottage gardens".

Modern-day cottage gardens include countless regional and personal variations of the more traditional English cottage garden,and embrace plant materials,such as ornamental grasses or native plants,that were never seen in the rural gardens of cottagers.Traditional roses,with their full fragrance and lush foliage,continue to be a cottage garden mainstay — along with modern disease-resistant varieties that keep the traditional attributes.Informal climbing plants,whether traditional or modern hybrids,are also a common cottage garden plant.Self-sowing annuals and freely spreading perennials continue to find a place in the modern cottage garden,just as they did in the traditional cottager's garden.

HISTORY

Origins

Fig. Vernacular thatched cottages (built in 1812–1816) in Woburn Street,Ampthill,Bedfordshire,surrounded by garden.

Cottage gardens,which emerged in Elizabethan times,appear to have originated as a local source for herbs and fruits.One theory is that they arose out of the Black Death of the 1340s,when the death of so many laborers made land available for small cottages with personal gardens.According to the late 19th-century legend of origin,these gardens were originally created by the workers that lived in the cottages of the villages,to provide them with food and herbs,with flowers planted in for decoration.Helen Leach analysed the historical origins of the romanticised cottage garden,subjecting the garden style to rigorous historical analysis,along with the ornamentalpotager and the herb garden.She concluded that their origins were less in workingmen's gardens in the 19th century and more in the leisured classes' discovery of simple hardy plants,in part through the writings of John Claudius Loudon.Loudon helped to design the estate at Great Tew,Oxfordshire,where farm workers were provided with cottages that had architectural quality set in a small garden—about an acre—where they could grow food and keep pigs and chickens.

Authentic gardens of the yeoman cottager would have included a beehive and livestock,and frequently a pig and sty,along with a well.The peasant cottager of medieval times was more interested in meat than flowers,with herbs grown for medicinal use and cooking,rather than for their beauty.By Elizabethan times there was more prosperity,and thus more room to grow flowers.Even the early cottage garden flowers typically had their practical use—violets were spread on the floor (for their pleasant scent and keeping out vermin); calendulas and primroses were both attractive and used in cooking.Others,such as sweet william and hollyhocks were grown entirely for their beauty.

Development

The "naturalness" of informal design began to be noticed and developed by the British leisured class.Alexander Pope was an early proponent of less formal gardens,calling in a 1713 article for gardens with the "amiable simplicity of unadorned nature".Other writers in the 18th century who encouraged less formal,and more natural,gardens includedJoseph Addison and Lord Shaftesbury.The evolution of cottage gardens can be followed in the issues of *The Cottage Gardener* (1848–61),edited by George William Johnson,where the emphasis is squarely on the "florist's flowers",carnations and auriculas in fancy varieties that were originally cultivated as a highly competitive blue-collar hobby.

Fig. Restored Gertrude Jekyll border at Manor House,Upton Grey,Hampshire.

William Robinson and Gertrude Jekyll helped to popularise less formal gardens in their many books and magazine articles.Robinson's*The Wild Garden*,published in 1870,contained in the first edition an essay on "The Garden of British Wild Flowers",which was eliminated from later editions.In his *The English Flower Garden*,illustrated with cottage gardens from Somerset,Kent and Surrey,he remarked,"One lesson of these little gardens,that are so pretty,is that one can get good effects from simple materials." From the 1890s his lifelong friend Jekyll applied cottage garden principles to more structured designs in even quite large country houses.Her *Colour in the Flower Garden* (1908) is still in print today.

Robinson and Jekyll were part of the Arts and Crafts Movement,a broader movement in art,architecture,and crafts during the late 19th century which advocated a return to the informal planting style derived as much from the Romantic tradition as from the actual English cottage garden.The Arts and Crafts Exhibition of 1888 began a movement toward an idealised natural country garden style.The garden designs of Robinson and Jekyll were often associated with Arts and Crafts style houses.Both were influenced by William Morris,one of the leaders of the Arts and Crafts Movement—Robinson quoted Morris's views condemning carpet bedding; Jekyll shared Morris's mystical view of nature and drew on the floral designs in his textiles for her gardening style.When Morris built his Red House in Kent,it influenced new ideas in architecture and gardening—the "old-fashioned" garden suddenly became a fashion accessory among the British artistic middle class,and the cottage garden esthetic began to emigrate to America.

Modern day

In the early 20th century the term "cottage garden" might be applied even to as large and sophisticated a garden as Hidcote Manor,which Vita Sackville-West described as "a cottage garden on the most glorified scale" but where the colour harmonies were carefully contrived and controlled,as in the famous "Red Borders".Sackville-West had taken similar models for her own "cottage garden",one of many "garden rooms" at Sissinghurst Castle—her idea of a cottage garden was a place where "the plants grow in a jumble,flowering shrubs mingled with Roses,herbaceous plants with bulbous subjects,climbers scrambling over hedges,seedlings coming up wherever they have chosen to sow themselves".The cottage garden ideal was also spread by artists such as water-colourist Helen Allingham (1848–1926).Another influence was Margery Fish (1892–1969),whose garden survives at East Lambrook Manor.

The cottage garden in France is a development of the early 20th century.Monet's garden at Giverny is a prominent example,a sprawling garden full of varied plantings,rich colors,and water gardens.In modern times,the term 'cottage garden' is used to describe any number of informal garden styles,using design and plants very different from their traditional English cottage garden origins.Examples include regional variations using a grass prairie scheme (in the American midwest) and California chaparral cottage gardens.

Design

While the classic cottage garden is built around a cottage,many cottage-style gardens are created around houses and even estates such as Hidcote Manor,with its more intimate "garden rooms".The cottage garden design is based more on principles than formulae: it has an informal look,with a seemingly casual mixture of flowers,herbs,and vegetables often packed into a small area.In spite of their appearances,cottage gardens have a design and formality that help give them their grace and charm.Due to space limitations,they are often in small rectangular plots,with practical functioning paths and hedges or fences.The plants,layout,and materials are chosen to give the impression of casualness and a country feel.Modern cottage gardens frequently use local flowers and materials,rather than those of the traditional cottage garden.What they share with the tradition is the unstudied look,the use of every square inch,and a rich variety of flowers,herbs,and vegetables.

The cottage garden is designed to appear artless,rather than contrived or pretentious.Instead of artistic curves,or grand geometry,there is an artfully designed irregularity.Borders can go right up to the house,lawns are replaced with tufts of grass or flowers,and beds can be as wide as needed.Instead of the discipline of large scale color schemes,there is the simplicity of harmonious color combinations between neighbouring plants.The overall appearance can be of "a vegetable garden that has been taken over by flowers." The method of planting closely packed plants was supposed to reduce the amount of weeding and watering required,but planted stone pathways or turf paths,and clipped hedges overgrown with wayward vines,are cottage garden features requiring well-timed maintenance.

Materials

Paths,arbors,and fences use traditional or antique looking materials.Wooden fences and gates,paths covered with locally made bricks or stone,and arbors using natural materials all give a more casual—and less formal—look and feel to a cottage garden.Pots,ornaments,and furniture also use natural looking materials with traditional finishes—everything is chosen to give the impression of an old-fashioned country garden.

Plants

Cottage garden plants are chosen for their old-fashioned and informal appeal.Many modern day gardeners use heirloom or 'old-fashioned' plants and varieties—even though these may not have been authentic or traditional cottage garden plants.In addition,there are modern varieties of flowers that fit into the cottage garden look.For example,modern roses developed by David Austin have been chosen for cottage gardens because of their old-fashioned look (multi-petaled form and rosette-shaped flowers) and fragrance—combined with modern virtues of hardiness,repeat blooming,and disease-resistance.Modern cottage gardens often use native plants and those adapted to the local climate,rather than trying to force traditional English plants to grow in an incompatible environment—though many of the old favorites thrive in cottage gardens throughout the world.

Roses

Fig.A climbing sport of the elite'Souvenir de la Malmaison',introduced before 1893,typical of a modern cottage garden.

Cottage gardens are always associated with roses: shrub roses,climbing roses,and old garden roses with lush foliage,in contrast to the gangly modern hybrid tea roses.Old cottage garden roses include cultivated forms of *Rosa gallica*,which form dense mounded shrubs 3–4 ft high and wide,with pale pink to purple flowers—with single form to full double form blooms.They are also very fragrant,and include the ancient Apothecary's rose (*R.gallica* 'Officinalis'),whose magenta flowers were preserved solely for their fragrance.Another old fragrant cottage garden rose is the Damask rose,which is still grown in Europe for use in perfumes.Cultivated forms of this grow 4 to 6 ft or higher,with gently arching canes that help give an informal look to a garden.Even taller generally are the Alba roses,which are not always white,and which bloom well even in partial shade.

The Provence rose or *Rosa centifolia* is the full and fat "cabbage rose" made famous by Dutch masters in their 17th-century paintings.These very fragrant shrub roses grow 5 ft tall and wide,with a floppy habit that is aided by training on an arch or pillar.The centifolia roses have produced many descendants that are also cottage garden favorites,including varieties of moss rose (roses with attractive 'mossy' growth on their flower stalks and flower buds).Unlike most modern hybrids,the older roses bloom on the previous year's wood,so they aren't pruned back severely each year.Also as they don't bloom continuously,they can share their branches with later-flowering climbers such as Clematis vines,which use the rose branches for support.A rose in the cottage garden is not segregated with other roses,with bare earth or mulch underneath',but is casually blended with other flowers,vines,and groundcover.

With the introduction of China roses (derived from *Rosa chinensis*) late in the 18th century,many hybrids were introduced that had the remontant (repeat-blooming) nature of the China roses,but maintained the informal old rose shape and flower.These included the Bourbon rose and the Noisette rose,which were added to the rose repertoire of the cottage garden,and,more recently,hybrid "English" roses introduced by David Austin.

Climbing plants

Fig.*Clematis vitalba*

Many of the old roses had cultivars that grew very long canes,which could be tied to trellises or against walls.These older varieties are called "ramblers",rather than "climbers".Climbing plants in the traditional cottage garden included European honeysuckle (*Lonicera periclymenum*) and Traveller's Joy (*Clematis vitalba*).The modern cottage garden includes many Clematis hybrids that have the old appeal,with sparse foliage that allows them to grow through roses and trees,and along fences and arbors.There are also many Clematis species used in the modern cottage garden,including *Clematis armandii*,*Clematis chrysocoma*,and *Clematis flammula*.Popular honeysuckles for cottage gardens include Japanese honeysuckle and *Lonicera tragophylla*.

Hedging plants

In the traditional cottage garden,hedges served as fences on the perimeter to keep out marauding livestock and for privacy,along with other practical uses.Hawthorn leaves made a tasty snack or tea,while the flowers were used for making wine.The fast-growingElderberry,in addition to creating a hedge,provided berries for food and wine,with the flowers being fried in batter or made into lotions and ointments.The wood had many uses,including toys,pegs,skewers,and fishing poles.Holly was another hedge plant,useful because it quickly spread and self-seeded.Privet was also a convenient and fast-growing hedge.Over time,more ornamental and less utilitarian plants became popular cottage garden hedges,including laurel,lilac,snowberry,japonica,and others.

Flowers and herbs

Fig. Lavender

Popular flowers in the traditional cottage garden included florist's flowers which were grown by enthusiasts—such as violets,pinks,and primroses—and those grown with a more practical purpose.For example,the calendula,grown today almost entirely for its bright orange flowers,was primarily valued for eating,for adding color to butter and cheese,for adding smoothness to soups and stews,and for all kinds of healing salves and preparations.Like many old cottage garden annuals and herbs,it freely self-sowed,making it easier to grow and share.Other popular cottage garden annuals included violets, pansies, stocks, and mignonette.

Perennials were the largest group of traditional cottage garden flowers—those with a long cottage garden history include hollyhocks,carnations,sweet williams, marguerites, marigolds, lilies, peonies, tulips, crocus, daisies, foxglove, monkshood, lavender, campanulas, Solomon's seal,evening primrose,lily-of-the-valley,primrose,cowslips,and many varieties of roses.

Today herbs are typically thought of as culinary plants,but in the traditional cottage garden they were considered to be any plant with household uses.Herbs were used for medicine,toiletries,and cleaning products.Scented herbs would be spread on the floor along with rushes to cover odors.Some herbs were used for dyeing fabrics.Traditional cottage garden herbs included sage,thyme,southernwood,wormwood,catmint,feverfew,lungwort,soapwort,hyssop,sweet woodruff,and lavender.

Fruits

Fruit in the traditional cottage garden would have included an apple and a pear,for cider and perry,gooseberries and raspberries.The modern cottage garden includes many varieties of ornamental fruit and nut trees,such as crabapple and hazel,along with non-traditional trees like dogwood.

7

Planting the Garden

The old saying that "patience is a virtue" applies to gardeners who get the itch to garden when temperatures warm up in the spring.One of the ways to determine when to plant veggies is based on their hardiness or their ability to withstand frost and cold temperatures.

Very hardy vegetables can be planted four to six weeks before the frost-free date in the spring.Potato tubers and onion sets can be planted. Asparagus,broccoli and cabbage can be planted as transplants. Collards, spinach, peas, lettuce and turnips can be planted from seed. Frost tolerant vegetables can be planted two to three weeks before the frost-free date.Cauliflower can be planted as a transplant. Carrots, mustard, parsnip, beets and radishes can be planted from seed. Tender vegetables can be planted on or after the frost-free date.Beans,sweet corn and summer squash can be planted from seed.Tomatoes transplants can be planted. Warm-loving vegetables can be planted one to two weeks after the frost-free date.Warm loving vegetables need warm temperatures and warm soil before planting. Vining crops like watermelon,cucumbers,pumpkins and cantaloupe can be planted. Pepper, eggplant and sweet potatoes should also be planted.

WHAT TO PLANT

Don't go overboard with your seed ordering after viewing all the colorful garden catalogs with their beautiful pictures of veggies or you may be the gardener in your neighborhood trying to give away zucchini.Grow what your family likes to eat.As a first time gardener,stay away from "exotic" veggies like kohlrabi or hard to grow veggies like cauliflower or head lettuce.

Grow hybrid vegetables.Hybrid vegetables are usually stronger and healthier than other vegetables.They often have higher yields.Many have a built-in disease resistance and they are more likely to recover from bad weather.Hybrids may cost a little bit more than other types of vegetables,but the cost is worth it.If you save seeds,remember that hybrids do not reproduce true to type,meaning the new plant will be inferior to the mother plant.

Choose vegetables that have earned the All-America Selections award.All-America Selections is an organization that has been evaluating new vegetable varieties in trialand display gardens across the United States and Canada since 1933.Each year after the evaluations have been analyzed a number of the most outstanding vegetables are designated as All-America Selections indicating that they performed well under all types of conditions.

Choose disease resistant varieties of vegetables.Just because a vegetable has built-in resistance to a specific disease doesn't mean that that vegetable will not get the disease,but it will fare better than a vegetable that has no resistance to the disease.Verticillium and fusarium wilt attack tomatoes.The tomato variety 'Celebrity' is resistant to both wilt diseases.The letters VF by the variety's name in a garden catalog or on a plant label indicates that the plant is resistant to both wilts.

HOW TO PLANT

After digging your soil to a depth of 6-10 inches,break up any large clods with a rake.Use the rake to prepare a smooth seedbed.Spread 1 1/2 pounds of a vegetable garden fertilizer over every 100 square feet of your vegetable garden.A one pound coffee can hold 1 1/2 pounds of fertilizer.Rake the fertilizer into the top 2-4 inches of soil. If you plan to use an organic fertilizer,add a two to four inch layer of organic matter over the vegetable garden and dig it into the soil.Organic matter will improve your soil structure besides adding nutrients to the soil.

Seeding

Before seeding,be sure you have created a smooth seedbed.To avoid compacting the soil,try to avoid walking over areas you will be seeding and planting.Be sure to follow the directions on the seed packet for planting depth of seeds.As a general rule seeds should be planted to a depth 2 to 4 times their diameter or largest width.Cover the seed with soil and tamp it down with the back of your hoe.Water lightly and keep moist until germination occurs.

Transplants

Vegetables like tomatoes,peppers,cabbage,broccoli,eggplants and collards are planted as transplants.

When you purchase transplants,choose transplants with the following characteristics.

- Choose plants with healthy green leaves.Avoid plants with yellowing or browning leaves.These plants may be diseased.
- Avoid plants in pots with roots growing out of the drainage hole.This usually indicates the plant may be root bound.Tap the plant out of the pot and check the roots.Healthy roots will be white.Roots that have browned are dead.Avoid purchasing these plants.

- Check the plants for insects.Shake the plant.If you see tiny white flying insects,they may be whiteflies.Check the undersides of the leaves for aphids.Aphids are tiny,oval shaped insects that cluster on the undersides of leaves.Both of these insects are sucking insects that will cause browning and curling of leaves.Do not buy insect infested plants.

Before planting your transplants,harden off your plants.Hardening off is a process of slowly introducing transplants to cooler temperatures and brighter light conditions outdoors.Gradually increase the time your transplants spend outdoors over a week to ten days before planting.

Try to plant on a cloudy day or in the late afternoon to avoid planting in high temperatures.Planting in high temperatures will put your plants under a lot of stress.Dig a hole big enough for the plant's root ball.Try not to damage the root system as you remove the plant from its pot.Space the transplants at recommended distances. Water your transplants in with a cup of a starter fertilizer.Mix one to two tablespoons of a soluble starter fertilizer with a gallon of water.A starter fertilizer is high in phosphorus. Phosphorus helps to promote root development. Promoting root development will get your plant off to a good start. Be sure to label all the plants in your garden.It is very difficult to identify plants,especially just after germination.

PLANTING TIPS

Save orange juice and tuna fish cans to use as barriers around newly transplanted plants to protect them from the cutworm.Cutworms will chew through the stems at soil level.Cut both ends from the cans and push cans about an inch into the soil around the plants.After two to three weeks,the cans can be removed because the stems will have thickened enough to withstand any cutworm damage.

Water vegetable transplants with a starter fertilizer.This should be water soluble,high phosphorus (N-P-K) mixed fertilizer.Phosphorus helps to promote root growth. Protect cucurbit crops (cucumbers,melons,squash,pumpkins) from cucumber beetles and the cucumber wilt that they spread as they feed with floating row covers after planting.Make sure to remove the row cover after the plants have begun flowering,so that they can be pollinated.

Tomatoes are subject to a few diseases.Verticillium and fusarium wilts are soil borne diseases that cause yellowing of the leaves,wilting and premature death of plants.Resistant varieties are the best preventative.Resistant tomatoes will have "VF" on the label. Carrots can be planted as early as the end of March/ first of April.To get long straight carrots the soil should be loose,worked deeply,well drained and have no clods or rocks in the soil. Plant onion sets in April.Buy sets early before they start sprouting in garden centers.Divide the sets up into those that are larger than a dime in diameter and those smaller.The bigger sets are best grown for green onions.The smaller sets make the best large onions for storage.Torpedo-shaped onions will produce round onions while the round sets will produce flat onions.For green onions,plant the bigger sets one inch deep and touching each other.For large,dry onions plant the small sets one inch deep and two to four inches apart.

Buy healthy vegetable transplants.Leaves and stems should be green and healthy without any signs of yellowing or browning.Yellowing or browning leaves may indicate an insect or disease problem.Gently remove transplants from their tray and check the root system.Roots should be white with visible soil.Transplants with brown dead roots should not be purchased.Check for insects such as whiteflies or aphids.Be sure to gradually introduce your transplants to the outdoor environment over a period of days,especially plants grown and purchased in a greenhouse.When you do plant,water your transplants in with a starter fertilizer that is high in phosphorus which helps to promote root development. Plant flowers in your vegetable garden.Many flowers will attract the beneficial insects,parasites and predators that help control pests.Good choices are sweet alyssum, dill, fennel, tansy, cosmos, yarrow, coneflower and sunflower. Choose disease resistant varieties.Provide good air circulation to help control disease.Stake or cage plants and allow proper spacing.

Time plantings to avoid insect problems.For instance,to avoid the worst time for squash vine borer and corn earworm,plant squash and corn so it can be harvested by July. Sow radish,lettuce,spinach,beet and turnip seed late in August.These vegetables will mature in the cooler fall weather.

VEGETABLE GARDENING FOR BEGINNERS

ORGANIC VEGETABLE GARDENS BRING SPECIAL REWARDS

Starting a vegetable garden? Dream big,but start small and expand as you gain experience.Raised beds make efficient use of space and keep maintenance to a minimum.

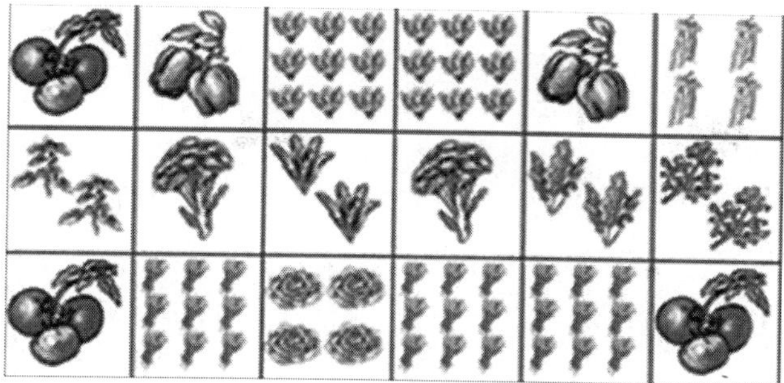

The Kitchen Garden Planner has a set of pre-planned gardens — a great starting point for beginners.You can also use the planner to design your own.

GROWING your own vegetables is both fun and rewarding.All you really need to get started is some decent soil and a few plants.But to be a really successful vegetable gardener — and to do it organically — you'll need to understand what it takes to keep your plants healthy and vigorous.Here are the basics.

"Feed the soil" is like a mantra for organic gardeners,and with good reason.In conventional chemical agriculture,crop plants are indeed "fed" directly using synthetic fertilizers.

When taken to extremes,this kind of chemical force-feeding can gradually impoverish the soil.And turn it from a rich entity teeming with microorganisms insects and other life forms,into an inert growing medium that exists mainly to anchor the plants' roots,and that provides little or no nutrition in its own right.

Although various fertilizers and mineral nutrients (agricultural lime,rock phosphate,greensand,etc.) should be added periodically to the organic garden,by far the most useful substance for building and maintaining a healthy,well-balanced soil is organic matter.You can add organic matter to your soil many different ways,such as compost,shredded leaves,animal manures or cover crops.

Organic matter improves the fertility,the structure and the tilth of all kinds of soils.In particular,organic matter provides a continuous source of nitrogen and other nutrients that plants need to grow.It also provides a rich food source for soil microbes.As organisms in the soil carry out the processes of decay and decomposition,they make these nutrients available to plants.

Make Efficient Use of Space

The location of your garden (the amount of sunlight it receives,proximity to a source of water,and protection from frost and wind) is important.Yet just as crucial for growing vegetables is making the most of your garden space.

Lots of people dream of having a huge vegetable garden,a sprawling site that will be big enough to grow everything they want,including space-hungry crops,such as corn,dried beans,pumpkins and winter squash,melons,cucumbers and watermelons.If you have the room and,even more importantly,the time and energy needed to grow a huge garden well,go for it.But vegetable gardens that make efficient use of growing space are much easier to care for,whether you're talking about a few containers on the patio or a 50-by-100-foot plot in the backyard.Raised beds are a good choice for beginners because they make the garden more manageable.

GET RID OF YOUR ROWS

Shop for Raised Beds

Make your own raised bed with our Raised Bed Corners,or choose a complete kit.Elevated raised beds allow for no-bend gardening.

The first way to maximize space in the garden is to convert from traditional row planting to 3- or 4-foot-wide raised beds.Single rows of crops,while they might be efficient on farms that use large machines for planting,cultivating,and harvesting,are often not the best way to go in the backyard vegetable garden.In a home-sized garden,the fewer rows you have,the fewer paths between rows you will need,and the more square footage you will have available for growing crops.

If you are already producing the amount of food you want in your existing row garden,then by switching to raised beds or open beds you will actually be able to downsize the garden.By freeing up this existing garden space,you can plant green-manure crops on the part of the garden that is not currently raising vegetables and/or rotate growing areas more easily from year to year.Or you might find that you now have room for planting new crops — rhubarb,asparagus,berries,or flowers for cutting — in the newly available space.

Other good reasons to convert from rows to an intensive garden system:

Less effort.When vegetables are planted intensively they shade and cool the ground below and require less watering,less weeding,less mulching — in other words,less drudgery for the gardener.

Less soil compaction.The more access you have between rows or beds,the more you and others will be compacting the soil by walking in them.By increasing the width of the growing beds and reducing the number of paths,you will have more growing area that you won't be walking on,and this untrammeled soil will be fluffier and better for plants' roots.

GROW UP,NOT OUT

Shop for Vegetable Supports

Get the right support for every vegetable.Tomato cages,trellises and more.Next to intensive planting,trellising represents the most efficient way to use space in the garden.People who have tiny gardens will want to grow as many crops as possible on vertical supports,and gardeners who have a lot of space will still need to lend physical support to some of their vegetables,such

Table. Planting dates and distances for garden vegetables

	Planting dates		Planting distances (in inches)			
Vegetable	**Start seed indoors**	**Plant seed or plant outdoors**	**Between rows, hand cultivated**	**Between plants**	**Depth of seeding (inches)**	**Amount to order per 20 feet of row "Packet" refers to average commercially-packaged seed packet.**
Asparagus		April 15 - May 1 (crowns)	36	12 - 18	6 - 8 (crowns)	15 crowns
Beans, snap (bush)		May 15 - July 1	18 - 24	3 - 4	1½ - 2	3 - 4 oz
Beans, snap(pole)		May 15 - July 1	36	4 - 6	1½ - 2	2 - 3 oz
Beans, dry shell		May 15	18 - 24	3 - 4	1½	3 - 4 oz
Beans, lima		May 15 - June 10	18 - 24	4 - 6	1½	3 - 4 oz
Beets		April 15 - July 1	12 - 18	2 - 4	½ - 1	1 packet
Broccoli	March 1 - 15	April 15 or June 1	24 - 30	24	¼ (indoors)	1 packet or 9 plants
Brussels sprouts	March 1 - 15	April 15 or June 1	24 - 30	24	¼ (indoors)	1 packet or 9 plants
Cabbage, early	March 1 - 15	April 1 - May 1	24 - 30	18	¼ (indoors)	1 packet or 12 plants
Cabbage, late	April 15 - May 1	June 1	24 - 30	24	¼ (seedbed)	1 packet or 9 plants
Cabbage, Chinese		July 1	24 - 30	18	½	1 packet
Carrots		April 15 - June 15	18 - 24	2 - 3	¼	1 packet
Cauliflower	March 1 - 15	April 15 or June 1	24 - 30	18 - 24	¼ (indoors)	1 packet or 12 plants
Celery	Feb. 15 - March 1	May 15	18 - 24	8	1/8(indoors)	1 packet or 24 plants
Chard, Swiss		May 1	18 - 24	6 - 8	1	1 packet
Collards		April 15	24 - 36	6	¼	1 packet
Cucumbers		May 1 - June 15	48 - 60	12 between single plants; 36 between hills of three	1	1 packet

Vincent A.Fritz,Extension Vegetable Specialist
Southern Research and Outreach Center,Waseca/Department of Horticultural Science,St.Paul

GARDENING WITH EPSOM SALT

Epsom salt is a popular and well-reputed supplement in organic gardening.With the recent push towards "green" living,Epsom salt is an ideal answer to a variety of organic gardening needs.Both cost effective and gentle on your greenery,Epsom salt is an affordable and green treatment for your well-tended plants—both indoors and out.Completely one-of-a-kind with a chemical structure unlike any other,Epsom salt (or Magnesium Sulfate) is one of the most economic and versatile salt-like substances in the world.Throughout time,Epsom salt has been known as a wonderful garden supplement,helping to create lush grass,full roses,and healthy,vibrant greenery.It has long been considered a planter's "secret" ingredient to a lovely,lush garden,and is such a simple,affordable way to have a dramatic impact.Just as gourmet salt works with the ingredients in food to enhance and bring a meal to its full potential,Epsom salt enhances fertilizer and soil's capabilities to bring a deeper level of vitality to your garden's composition.Ultra Epsom Salt is the highest quality Epsom salt available,and is widely celebrated for its powerful benefits on natural life,ranging from household plants to shrubs,lawns and even trees.

Why Epsom Salt Works in the Garden

Composed almost exclusively of Magnesium Sulfate,Epsom salt is intensely rich in these two minerals that are both crucial to healthy plant life.These same minerals which are so beneficial for bathing and using around the house are also a wonderful facilitator to your garden,helping it reach its fullest potential and creating a lush and vibrant outdoor space.Unlike common fertilizers,Epsom Salt does not build up in the soil over time,so it is very safe to use.

MAGNESIUM

Magnesium is beneficial to plants from the beginning of their life,right when the seed begins to develop.It assists with the process of seed germination; infusing the seed with this important mineral and helping to strengthen the plant cell walls,so that the plant can receive essential nutrients.Magnesium also plays a crucial role in photosynthesis by assisting with the creation of chlorophyll,used by plants to convert sunlight into food.In addition,it is a wonderful help in allowing the plant to soak up phosphorus and nitrogen,which serve as vital fertilizer components for the soil.Magnesium is believed to bring more flowers and fruit to your garden,increasing the bounty as well as the beauty of your space.

SULFATE

Sulfate,a mineral form of sulfur found in nature,is an equally important nutrient for plant life.Sulfate is essential to the health and longevity of plants,and aides in the production of chlorophyll.It joins with the soil to make key nutrients more effective for plants,including nitrogen,phosphorus and potassium.Sulfate works in conjunction with Magnesium to create a “vitamin” full of minerals,nourishment and health benefits for your garden.

HOW TO USE EPSOM SALT IN THE GARDEN

EPSOM SALT FOR HOUSEPLANTS

Perhaps the most natural and easiest place to start with Ultra Epsom Salt is with the potted plants that are dispersed around your house and porch.Epsom salt is such a simple way to increase their blooming and health,and is something that you can include easily as a part of a normal routine.For potted plants,simply dissolve 2 tablespoons per gallon of water,and substitute this solution for normal watering at least once a month – although it is safe to do this as often as desired.

Adding this Epsom salt solution to houseplants that have been potted for a long time is especially useful,due in part to natural salt,which can build up in the soil and clog the root cells of the plant.Ultra Epsom Salt can help to clear up this accumulation of natural salts in the pot,and lead to a revival in the plant's health and vibrancy.

It is also useful for a plant that has just been potted,as it will more easily receive the proper nutrients and have a healthy start in life.As general guidance,most plants need plenty of sun to receive the benefits of Ultra Epsom Salt (and photosynthesize),so be sure to keep typical houseplants in a sunny area of the home unless instructed otherwise.Using Ultra Epsom Salt with potted vegetable plants is a really wonderful idea as well,because it can increase the amount of fruit or vegetables you receive from the one plant.This is particularly beneficial to apartment dwellers and those with little or no personal yard space,as Ultra Epsom Salt can help you receive a large bounty within a confined space.A wonderful way to easily and effectively grow food!

FIRST PLANTING WITH EPSOM SALT

For setting up your garden and the initial planting stage,Ultra Epsom Salt is especially useful for getting a nourishing start.Prep your garden soil by sprinkling up to 1 cup of Ultra Epsom Salt per 100 square feet,and then work it into the soil before seeding or planting.

This helps the seeds to germinate better,and start with a strong and healthy growth.It is also very beneficial for more mature plants that you are going to add to your garden,since the transition can be difficult for their growth and health.

VEGETABLE GARDENS & EPSOM SALT

For maintaining and creating a vegetable garden,Epsom salt can help you refresh and revitalize the garden you have already created—or create a healthy beginning to a new space.Ultra Epsom Salt is advised for use with all fruits,vegetables,and herbs *(It is not advisable to use Epsom salt with the planting of sage—it is not beneficial for this particular plant)*.As previously mentioned,it does not cause build-up or any harm to plants when used,and so can be used safely and effectively during any stage of the plant's life.For general purposes,Ultra Epsom Salt works well as a saline solution for a tank sprayer.Simply fill your tank sprayer (commonly available at gardening and home improvement stores) with 1 tablespoon of Ultra Epsom Salt per gallon of water.Then spray your garden after the initial planting,later when it begins to grow (or after a month or so for transplants),and lastly when the vegetables begin to mature.It is believed that this practice will give you healthier vegetables,and a lush vegetable garden.

The advice above is wonderful for any vegetable or herb,but we do have additional advice for some varieties and situations:

Tomatoes & Epsom Salt

Tomatoes are prone to magnesium deficiency later in the growing season,and display this through yellow leaves and less production.They can greatly benefit from Ultra Epsom Salt treatments both at the beginning of their planting and throughout their seasonal life.When gardening,simply add one or two tablespoons per hole before planting the seeds or transplants.Then as the tomato matures,either work in one tablespoon of Ultra Epsom Salt per foot of plant height around the base of the tomato plant (individually),or create the tank sprayer solution mentioned above and use that every two weeks.

Peppers & Epsom Salt

Like tomatoes,peppers are also prone to magnesium deficiency and thrive much more fully with the use of Epsom salt.This can be done in the same way as tomatoes—through adding one or two tablespoons per hole before planting (for seeds and grown plants),and then twice a week based on the height of the plant.A study conducted by the National Gardening Association discovered that four out of six home gardeners noticed that their Epsom salt-treated peppers were larger than those that were un-treated.

Many gardeners credit their healthy,vibrant peppers and tomatoes to Epsom salt.This solution truly aides in the production level,aesthetic beauty and quality of the harvest produced.

FLOWER GARDENS & EPSOM SALT

Like vegetable gardens,flower gardens also blossom more vibrantly and beautifully with the use of Ultra Epsom Salt in the soil and as a liquid solution.Epsom salt helps your garden to become the calming,serene environment you have been envisioning,and will increase the beauty of your home and landscape as well.To use,follow the guidelines outlined in the First Plantingsection for both brand new seedlings and more mature plants.Next,using a tank sprayer,fill with a liquid solution containing one tablespoon of Ultra Epsom Salt per gallon of water.This solution can be used as much as desired during the gardening season; but definitely after the initial planting,then later when you see growth (or after a month or so for transplants),and finally when they have received full bloom.If you don't have a tank sprayer,you can always create this solution in a watering can using the ratio of 1 tablespoon Ultra Epsom Salt to 1 gallon of water.

Roses and flower bushes have some additional tips concerning the use of Epsom salt:

Roses & Epsom Salt

Roses in particular can greatly benefit from Epsom salt,and it is said to make foliage greener,healthier and lead to more canes and roses.Start by soaking unplanted rose bushes in one half cup of Ultra Epsom Salt per gallon of water before planting,to help the roots get stronger and firmer.Then,when planting,add one tablespoon of Ultra Epsom Salt per hole before inserting the rose bush.After the roses are planted (and to boost already planted roses),make the liquid Ultra Epsom Salt solution listed above for either a tank sprayer or watering can,or simply work in one tablespoon of Ultra Epsom Salt per foot of plant (individually).Once during the beginning of the season,it is also advised to work one half cup of Ultra Epsom Salt into the base of the plant to encourage blooming canes and healthy basal cane development.

Shrubs & Epsom Salt

For flowering and green shrubs,particularly evergreens,azaleas and rhododendrons,Epsom salt can improve the blooming of the flowers and the vibrancy of the greenery.Simply work in one tablespoon of Ultra Epsom Salt per nine square feet of bush into the soil,over the root zone,which allows the shrubs to absorb the nutritional benefits.Repeat this every two to four weeks for optimal results.

LAWN CARE & EPSOM SALT

Just as Ultra Epsom Salt can revitalize your garden,so does it improve the greenery and sustainability of your lawn.Epsom salt is particularly useful for preventing a yellowing lawn and creating lusher,softer,deeply green grass.It can be applied using a tank sprayer (which can also be used on your flower and vegetable gardens),a lawn spreader,and by using a hose and spray attachment.Use three pounds per 1250 square feet (25' x 50'),six pounds per 2500 square feet (50' x 50'),and twelve pounds per 5000 square feet (50' x 100').If using a tank sprayer or a hose and spray attachment,make sure to dilute the salt in plenty of water (enough to make it dissolve),so that it is a concentrated solution.

TREES & EPSOM SALT

Trees,the largest and longest standing part of your garden,can also benefit from Epsom salt by allowing more minerals to be absorbed through the roots,giving you strong healthy trees to enjoy for years to come.If your trees bloom or produce fruit,Ultra Epsom Salt can be particularly useful due to its ability to increase the production of both flowers and bounty.Simply work in two tablespoons per nine square feet into the soil over the root zone three or four times a year.Planning to complete this at the beginning of each season is particularly helpful for preparing the tree for the change in weather,and allowing them to become stronger and healthier.

8

Kitchen Gardening

The traditional kitchen garden,also known as a potager (in French,*jardin potager*) or in Scotland a kailyaird,is a space separate from the rest of the residential garden – the ornamental plants and lawn areas.Most vegetable gardens are still miniature versions of old family farm plots,but the kitchen garden is different not only in its history,but also its design. The kitchen garden may serve as the central feature of an ornamental,all-season landscape,or it may be little more than a humblevegetable plot.It is a source of herbs,vegetables and fruits,but it is often also a structured garden space with a design based on repetitive geometric patterns. The kitchen garden has year-round visual appeal and can incorporate permanent perennials or woody shrub plantings around (or among) the annuals.

POTAGER GARDEN

A potager is a French term for an ornamental vegetable or kitchen garden.The historical design precedent is from the Gardens of the French Renaissance and Baroque Garden à la française eras.Often flowers (edible and non-edible) and herbs are planted with the vegetables to enhance the garden's beauty.The goal is to make the function of providing food aesthetically pleasing. Plants are chosen as much for their functionality as for their color and form.Many are trained to grow upward.A well-designed potager can provide food,as well as cut flowers and herbs for the home with very little maintenance.Potagers can disguise their function of providing for a home in a wide array of forms—from the carefree style of the cottage garden to the formality of a knot garden.

VEGETABLE GARDEN

A vegetable garden (also known as a vegetable patch or vegetable plot) is a garden that exists to grow vegetables and other plants useful for human consumption,in contrast to a flower garden that exists for aesthetic purposes.It is a small-scale form of vegetable growing.A vegetable garden typically includes a compost heap,and several plots or divided areas of land,intended to grow one or two types of plant in each plot.Plots may also be divided into rows with an assortment of vegetables grown in the different rows.It is usually located to the rear of a property in the back garden or back yard.Many families have home kitchen and vegetable gardens that they use to produce food.In World War II,many people had a garden called a "victory garden" which provided food and thus freed resources for the war effort.

Fig.A small vegetable garden in May outside of Austin,Texas

Fig. An herbal garden at Lippensgoed-Bulskampveld,Beernem,Belgium

Fig. Borage is commonly grown in herb gardens; its flowers can be used as agarnish

Fig. Cowbridge Physic Garden,Wales

With worsening economic conditions and increased interest in organic and sustainable living,many people are turning to vegetable gardening as a supplement to their family's diet.Food grown in the back yard consumes little if any fuel for shipping or maintenance,and the grower can be sure of what exactly was used to grow it.Organic horticulture,or organic gardening,has become increasingly popular for the modern home gardener.

There are many types of vegetable gardens.The potager,a garden in which vegetables,herbs and flowers are grown together,has become more popular than the more traditional rows or blocks.

HERB GARDEN

The herb garden is often a separate space in the garden,devoted to growing a specific group of plants known as herbs.These gardens may be informal patches of plants,or they may be carefully designed,even to the point of arranging and clipping the plants to form specific patterns,as in a knot garden.

Herb gardens may be purely functional or they may include a blend of functional and ornamental plants.The herbs are usually used to flavour food in cooking,though they may also be used in other ways,such as discouraging pests,providing pleasant scents,or serving medicinal purposes (such as a physic garden),among others.

A kitchen garden can be created by planting different herbs in pots or containers,with the added benefit of mobility.Although not all herbs thrive in pots or containers,some herbs do better than others.Mint,a fragrant yet invasive herb,is an example of an herb that is advisable to keep in a container or it will take over the whole garden. Some popular culinary herbs in temperate climates are to a large extent still the same as in the medieval period.

Herbs often have multiple uses.For example,mint may be used for cooking,tea,and pest control.Examples of herbs and their uses (not intended to be complete):

- Annual culinary herbs: basil,dill,summer savory
- Perennial culinary herbs: mint,rosemary,thyme,tarragon
- Herbs used for potpourri: lavender,lemon verbena
- Herbs used for tea: mint,lemon verbena,chamomile,bergamot,hibiscus
- Herbs used for other purposes: stevia for sweetening,feverfew for pest control in the garden.

Witches' garden

A witches' garden is an herb garden specifically designed and used for the cultivation of herbs,for culinary,medicinal and/or spiritual purposes.Herbal baths,the making of incense,tied in bundles for rituals or prayers,or placed in charms are just some of the ways herbs can be used for spiritual purposes.

Herb gardens developed from the general gardens of the ancient classical world,which were used for growing vegetables,flowers,fruits and medicines.For centuries "wise women" and "healers" understood the uses of herbs for the purposes of healing and magic.During the medieval period,monks and nuns acquired this medical knowledge and grew the necessary herbs in specialized gardens.

Typical plants found within a witches' garden are the following: rosemary, sage, parsley, mint, catnip, henbane, marjoram, thyme, rue, angelica, bay, oregano, dill,aloe,arnica,chives,and basil.Basil is especially common in these gardens,not just for its culinary use,but as a strong protection herb.It is said,"Where basil grows,no evil goes!" and "Where basil is,no evil lives!" With the advance of medical and botanical sciences in Renaissance Europe,monastic herb gardens developed into botanical gardens.However,these are just examples of the common herbs found within a witch's garden.Many other plants and herbs can be grown.It is a very personal garden,and therefore is unique to the individual witch.For a "true" witch or pagan,this garden is not just used for the purposes of each of the herbs grown,but it is also a way to become in touch with mother nature and become one with the Earth.

Starting a Kitchen Garden

If you have to choose between a sunny spot or a close one,pick the sunny one.The best location for a new garden is one receiving full sun (at least six hours of direct sunlight per day),and one where the soil drains well.If no puddles remain a few hours after a good rain,you know your site drains well.

After you've figured out where the sun shines longest and strongest,your next task will be to define your kitchen garden goals.My first recommendation for new gardeners is to start small,tuck a few successes under your belt in year one,and scale up little by little.

But what if you're really fired up about it? Even in year one,you may be able to meet a big chunk of your family's produce needs.In the case of my garden in Scarborough,Maine,we have 1,500 square feet under cultivation,which yields enough to meet nearly half of my family of five's produce needs for the year.When you do the garden math,it comes out to 300 square feet per person.More talented gardeners with more generous soils and climates are able to produce more food in less space,but maximizing production is not our only goal.We're also trying to maximize pleasure and health,both our own and that of the garden.Kitchen gardens and gardeners thrive because of positive feedback loops.If your garden harvests taste good and make you feel good,you will feel more motivated to keep on growing.

PREPARING THE GARDEN SITE

If you're starting your kitchen garden on a patch of lawn,you can build up from the ground with raised beds,or plant directly in the ground.Building raised beds is a good idea if your soil is poor or doesn't drain well,and you like the look of containers made from wood,stone or corrugated metal.This approach is usually more expensive,however,and requires more initial work than planting in the ground.

Whether you're going with raised beds or planting directly in the ground,you'll need to decide what to do with the sod.You can remove it and compost it,which is hard work,but ensures that you won't have grass and weeds coming up in your garden.If you're looking to start a small or medium-sized garden,it's possible to cut and remove sod in neat strips using nothing more than a sharp spade and some back muscle.For removing grass from a larger area,consider renting a sod cutter.

Otherwise,to avoid sod removal altogether,you can use a technique called "sheet mulching," or "lasagna gardening," whereby you smother the grass with one or more layers of organic goodness (untreated cardboard, newspaper, loam, compost, leaves, grass clippings,etc.) in such a way that the grass underneath dies and decomposes,enriching the soil with organic matter.Sheet mulching is particularly effective for kitchen gardens started in the fall,as the sod has more of a chance to break down over the winter.One variation on sheet mulching is to use instant garden beds,which are made by laying out bags of topsoil to form beds.

Choosing Garden Crops

The most important recommendation after "start small" is "start with what you like to eat." This may go without saying,but I have seen first-year gardens that don't reflect the eating habits of their growers — a recipe for disappointment.That said,I believe in experimenting with one or two new crops per year that aren't necessarily favorites for the sake of having diversity in the garden and on our plates. One of the easiest and most rewarding kitchen gardens is a simple salad garden.Lettuces and other greens don't require much space or maintenance,and grow quickly.Consequently,they can produce multiple harvests in most parts of the country.If you plant a "cut-and-come-again" salad mix,you can grow five to 10 different salad varieties in a single row.And if you construct a cold frame (which can be cheap and easy if you use salvaged storm windows),you can grow some hearty salad greens year-round.

When it comes to natural flavor enhancers,nothing beats culinary herbs.Every year I grow standbys such as parsley, chives, sage, basil, tarragon, mint, rosemary and thyme,but I also make an effort to try one or two new ones.One consequence of this approach is that I end up expanding my garden a little bit each year,but that's OK,because my skills and gastronomy are expanding in equal measure,as are my sense of satisfaction and food security.

Planting a Garden: Where,When and How

Next,sketch out a garden plan of what will be planted where,when and how.To do this,you need to get familiar with the various edible crops and what they like in terms of space,water,soil fertility and soil temperatures.

Garden Seeds or Transplants?

When the time comes to plant your kitchen garden,you'll need to decide which plants to start from seed and which to buy as transplants.Many gardeners choose to plant all of their crops from seed for a variety of reasons,including lower costs,greater selection,and the challenge and satisfaction of seeing a plant go from seed to soup bowl.But whether you're a greenhorn or a green thumb,there's no shame in buying seedlings.Doing so increases your chances of success,especially with crops such as eggplants,peppers and tomatoes that require a long growing season.(Learn more about transplanting.)

Mulch,Mulch...And More Mulch

After you've sown your seeds or planted your plants,introduce yourself to the kitchen gardener's best friend,Mr.Mulch.Just about any organic matter you can get your hands on — straw,grass clippings,pine needles,shredded leaves,dead weeds that haven't gone to seed — can be used as mulch.I bring in mulch from neighbors who would otherwise throw it away.Mulch plays three main roles: It deters weeds,helps retain moisture,and adds organic matter to the soil as it decays.I apply it to the pathways between my beds and around all of my plants.(Learn more about building healthy soil.)

When and How Much to Water Your Garden

Fruits and vegetables are made mostly of water,so you'll need to make sure your plants are getting enough to drink.This is especially important for seedlings that haven't developed a deep root structure.You'll want to water them lightly every day or two.Once the crops are maturing,they need about an inch of water per week,and more in sandy soils or hot regions.If Mother Nature isn't providing that amount of rain,you'll need to water manually or with a drip irrigation system.

Garden Maintenance: Keep An Eye On It

Sun and rain willing,fast growers such as radishes and salad greens will begin to produce crops as early as 20 to 30 days after planting.Check on them regularly so you get to harvest them before someone else does.In my garden,those "someones" include everything from the tiniest of bacteria to the largest of raccoons.Various protective barriers and organic products can deter pests and diseases,and if you have trouble with rabbits,deer or other four-legged critters,your best defense may be agarden fence.

Succession Planting: Plant Now and Later

Getting the most pleasure and production from your garden comes from learning the beauty of succession planting.Rather than trying to "get your garden in" during one busy weekend,space your planting out over the course of several weeks by using short rows.Every time you harvest a row or pull one out that has stopped producing,try to plant a new one.Succession plantings lead to succession harvests spread out over several months — one of the key characteristics of a kitchen garden.

As you gain new confidence and skills,you can look for ways to incorporate perennials including asparagus and rhubarb into your edible landscape.And no discussion of kitchen gardens would be complete without mentioning flowers,which should be added from the start.Flowers add beauty and color to the garden and the kitchen table.They also attractbeneficial insects while,in some cases,repelling undesirable ones.

Bibliography

Bradley,Richard: *New Improvements of Planting and Gardening.3d ed.London*: Mears,1719 and 1720.

Campbell,Susan.Charleston Kedding: *A History of Kitchen Gardening.London:* Ebury,1996.

Campbell,Susan.Cottesbrooke: *An English Kitchen Garden.London*: Century,1987.

Campbell,Susan: *Walled Kitchen Gardens.* Princes Risborough: Shire,1998.

Carter,George,Patrick Goode,and Kedrun Laurie,eds:*In the Catalogue for the Exhibition: Humphry Repton Landscape Gardener,1752–1818.Norwich:* Sainsbury Centre for Visual Arts,1982.

Davies,Jennifer: *The Victorian Kitchen Garden.London:* BBC Books,1987.

Evelyn,John: *The Compleat Gard'ner.Translated from the French Instructions pour les jardins frutiers et potagers by Jean-Baptiste de la Quintinye,1690.*London: Gillyflower,1693.

Howitt,William: *Rural Life of England.3rd ed.London: Longmans,Brown,Green,and Longmans*,1844.

Loudon,John Claudius: *Encyclopaedia of Gardening.5th ed.London: Longman, Rees, Orme, Brown, and* Green,1835.

M'Intosh,Charles: *The Book of the Garden.2 vols.* Edinburgh,1853–1855.

Morgan,Joan,and Alison Richards: *A Paradise Out of a Common Field.London*: Century,1990.

Mountain,Dydymus (alias Thomas Hill): *The Gardener's Labyrinth.New York.London: Garland*,1982.Facsimile of 1577.

Svieking,Alber Forbes,ed: Sir *William Temple upon the Gardens of Epicurus with Other Seventeenth-Century Essays*.Gollancz,1902.

Thomson,Robert: *The Gardener's Assistant,Practical and Scientific.1st ed.6 vols*.Glasgow,1859.

Trusler,Dr.John (attributed):*Elements of Modern Gardening: or,the Art of Laying Out of*

Pleasure Grounds,Ornamenting Farms,and Embellishing Views Round about Our Houses.London: Logographic Press,1784.

Wilson,C.Anne,ed.*The Country House Kitchen Garden,1600–1950.London*: Sutton Publishing with the National Trust,1998.

Adams,William Howard: *The French Garden,1500–1800*.New York: Braziller,1979.

Berger,Robert W: *In the Garden of the Sun King: Studies on the Park of Versailles Under Louis XIV. Washington,* D.C.: Dumbarton Oaks Research Library and Collection,1985.

Chambers,Douglas: *The Planters of the English Landscape Garden: Botany,Trees,and the Georgics.New Haven,Conn.,and London:* Yale University Press,1993.

Chase,Isabella Wakelin Urban.*Horace Walpole: Gardenist.An Edition of Walpole's "The History of the Modern Taste in Gardening," with an Estimate of Walpole's Contribution to Landscape Architecture*.Princeton,N.J.: Princeton University Press,1943.

Clunas,Craig: *Fruitful Sites: Garden Culture in Ming Dynasty China.Durham,N.C.*: Duke University Press,1996.

Taank, Praveen: *Advances in Forestry Research in India*, Cyber Tech Pub, Delhi, 2009.

Barrows, M.: *A Survey of the Intestinal Parasites of the Primates in Budongo Forest*, Uganda. Glasgow University, 1996.

Chakrabarty, Falguni: *Adaptation of the Santals to the Hill-Forest Environment*, APH, Delhi, 2012.

Gilbert Wooding Robinson: *Soils : Their Origin Constitution and Classification*, Biotech Books, Delhi, 2005.

Gupta, O.P. : *Water in Relation to Soils and Plants : With Special Reference to Agriculture*, Agrobios, Delhi, 2002.

Herminie Broedel Kitchen: *Soils and Crops : Diagnostic Techniques*, Satish Serial Publishing, Allahabad, 2004.

Owiunji, I.: *The Long Term Effects of Forest Management on the Bird Community of Budongo Forest Reserve*, Uganda. M. Sc., Makerere University, 1996.

Prasad, T.V.S. : *Soil Chemistry : Nutrient and Water Management in Agricultural Soils*, Dominant Pub, Delhi, 2009.

Reynolds, V.: *Budongo: A Forest and its Chimpanzees*, London, Methuen, 1965.

Index

L

M

N

O

P

Q

R

S

T

V

W